STATISTICS

STATISTICS

SECOND EDITION OF 'A SECOND COURSE IN STATISTICS'

ROBERT LOVEDAY

M.Sc. (Sheffield), F.I.S., F.I.M.A.

CAMBRIDGE

AT THE UNIVERSITY PRESS

1971

CAMBRIDGE
UNIVERSITY PRESS

University Printing House, Cambridge CB2 8BS, United Kingdom

Cambridge University Press is part of the University of Cambridge.

It furthers the University's mission by disseminating knowledge in the pursuit of education, learning and research at the highest international levels of excellence.

www.cambridge.org
Information on this title: www.cambridge.org/9781316606940

First edition 1961
Reprinted 1965, 1966
Second edition [metric] 1969
Reprinted 1971
First paperback edition 2016

A catalogue record for this publication is available from the British Library

ISBN 978-1-316-60694-0 Paperback

CONTENTS

PREFACE

The object of this volume is to establish firmly the bridge between the elementary treatment given in *A First Course in Statistics* and the rigorous treatment given in more advanced University courses.

The keyword of *A First Course in Statistics* is *observation.* The uninitiated student is unable to extract ideas from Statistics until he has learnt the common methods of classifying and representing data. He has to learn *what meaning* can be attached to the terms commonly used in Statistical Analysis. In a first course the idea of significance should be avoided and all differences at this stage should be absolutely blatant.

In this course, however, the important idea is *probability.* At this stage the student must be able to decide how much *confidence* he can place in his results; whether the small differences he observes are *significant* or not.

The opening chapter on location and dispersion makes it possible to begin this course without having previously read the *First Course.* In other words, this volume is complete in itself. The normal distribution is introduced in chapter 2 because experience shows that, immediately after they have become acquainted with *frequency distributions,* students feel an urgent need to know something about the *mathematical model* which fits so many of them so well. The brief treatment of probability in chapter 3, which includes Bayes' theorem, is sufficient to lead naturally to probability distributions in general and to the binomial and Poisson distributions in particular.

Chapters 7, 8 and 9 deal with the use of χ^2 and t-tables for testing significance and a short account of Quality Control follows in chapter 10. A treatment of regression by the method of least squares and of correlation coefficients (including dichotomy, Spearman's ρ and Kendall's τ) is given in chapters 10–13 and the final chapter covers the use of F-tables for the analysis of variance.

The glossary at the end of the book is of interest in that it not only summarises the terms and formulae introduced in the earlier part of the text but it also extends them. Thus, by reference to the glossary, a forward look can be taken to ideas left open for exploration in a more advanced treatise.

1 July 1969 R.L.

1

LOCATION AND DISPERSION

1. Location or central tendency. Fifteen 1968 British 10p coins, all completely new, were weighed separately, in grams. Their weights, in order, were:

12.48	12.63	12.64	12.65	12.66	12.67	12.68	12.71
12.73	12.77	12.78	12.80	12.82	12.90	12.93	

The *arithmetic mean,* or more simply the *mean,* of these fifteen weights is

$$\tfrac{1}{15}(12.48+12.63+12.64+\ldots+12.90+12.93) = \tfrac{1}{15}(190.85)\text{ g}$$
$$= 12.723\text{ g (3 decimals).}$$

Note that means are usually stated to one place (sometimes two places) of decimals more than the data.

The weight 12.723 g is commonly called the *average* of the fifteen weights, but in the study of Statistics it is usually called the mean. The mean of a group of observations is a measure of location of the group. It is a single number which enables us to assess the position in which the group is located with respect to other groups. This single number is also called the *central tendency* of the group.

Another useful measure of location is the *median.* This is the central observation of the group. In the above example it is 12.71 g. It has an equal number of observations above and below it and provides us with an actual specimen for the central tendency of the group. In a group of boys, the boy of median age can be interviewed, the physique of the boy of median weight can be examined, and the script of the examination candidate with the median mark can be subjected to further scrutiny. If a group contains an *even* number of observations the mean of the two central observations is taken as the median. The median is unaffected by abnormal individuals. It is also unaffected if the observations are *transformed* by some process such as *taking logarithms* or *squaring*. It is, however, unsuitable for work demanding mathematical manipulation.

Returning, therefore, to a study of the mean, we may regard the above example as a particular case of the general definition:

The mean of the n observations $x_1, x_2, \ldots, x_n$ *is*

$$\bar{x} = \frac{1}{n}(x_1 + x_2 + \ldots + x_n)$$

$$= \frac{1}{n}\sum_{r=1}^{n} x_r$$

and we may write

$$\bar{x} = \frac{1}{n}\Sigma x$$

if no confusion is likely to arise from the omission of the suffix and limits.

In 1968 the 10p cupro-nickel piece replaced the *florin*. Prior to 1947 the florin was minted from a *half silver* alloy. It is interesting, therefore, to compare the fifteen 10p cupro-nickel pieces with fifteen 'silver' florins minted between 1926 and 1946. When these were weighed, in grammes, their weights in order were:

12.05	12.05	12.08	12.15	12.60	12.70	12.71	12.72
12.83	12.86	12.89	12.90	12.91	12.92	12.93	

Thus for these fifteen florins, $\Sigma x = 189.30$ g and the mean weight $\bar{x} =$ 12.62 g exactly. This indicates how the weights of the florins are located with respect to the weights of the new 10p pieces. It is important to decide, however, if the difference between the means 12.723 g and 12.62 g is so small as to be *probably negligible* or if it is large enough to support the view that, on the average, new 10p pieces are heavier than used florins. In other words, 'Is the difference between the means *significant*?' This question is introduced in this opening section but not answered. It is one of the many examples of significance which will be discussed later.

2. Change of origin and unit. The calculation of the mean weight of the fifteen florins can be simplified by working with an *arbitrary origin* such as 12.40 g and at the same time taking 0.01 g as *unit*. The fifteen observations then become

−35	−35	−32	−25	20	30	31	32
43	46	49	50	51	52	53	

and their mean is 22. From this the mean weight of the fifteen florins is deduced as

$$12.40 + 0.01 \times 22 = 12.62 \text{ g.}$$

In general, if the n observations $x_1, x_2, \ldots, x_n$ are converted to $X_1, X_2, \ldots, X_n$ by working with A as an arbitrary origin and B as unit then

$$x_1 = A + BX_1, \quad x_2 = A + BX_2, \ldots, x_n = A + BX_n.$$

Hence

$$\Sigma x = nA + B\Sigma X \quad \text{and} \quad \bar{x} = A + B\bar{X}.$$

3. Variance. Standard deviation. If $\bar{x}$ is the mean of the n observations $x_1, x_2, \ldots, x_n$ then the n values

$$(x_1-\bar{x}), \quad (x_2-\bar{x}), \ldots, (x_n-\bar{x})$$

are called the *deviations from the mean*, and

$$(x_1-\bar{x})^2+(x_2-\bar{x})^2+\ldots+(x_n-\bar{x})^2$$

is the *sum of the squares* of these deviations.

The *variance* of the n observations $x_1, x_2, \ldots, x_n$ is defined as

$$\frac{1}{n}\{(x_1-\bar{x})^2+(x_2-\bar{x})^2+\ldots+(x_n-\bar{x})^2\} = \frac{1}{n}\Sigma(x-\bar{x})^2.$$

It may be described as the *mean-square deviation from the mean.* The *standard deviation* of the n observations $x_1, x_2, \ldots, x_n$ is

$$S = \sqrt{\left\{\frac{\Sigma(x-\bar{x})^2}{n}\right\}}.$$

It is the positive square root of the variance and may be described as the *root-mean-square deviation from the mean.* In the examples under discussion the standard deviation will be denoted by the large capital English letter S. In certain circumstances, to be discussed later, it will be denoted by the Greek letter σ or by the small English letter s.

The variance of the weights of the fifteen florins is, therefore,

$$S^2 = \tfrac{1}{15}\{(12.05-12.62)^2+(12.05-12.62)^2+\ldots+(12.93-12.62)^2\}\ \mathrm{g}^2$$
$$= 0.11403\ \mathrm{g}^2.$$

Note that:

(i) the units of this variance are g^2,

(ii) the above method of calculation is awkward,

(iii) if, as in the case of the 10p pieces the mean is not exact, there is an error in each of the fifteen deviations from the mean.

The difficulties (ii) and (iii) above can be avoided by realizing that the sum of the squares of deviations from the mean

$$\begin{aligned}\Sigma(x-\bar{x})^2 &= \Sigma(x^2-2x\bar{x}+\bar{x}^2)\\ &= \Sigma x^2-2\bar{x}\Sigma x+n\bar{x}^2\\ &= \Sigma x^2\times 2\bar{x}(n\bar{x})+n\bar{x}^2\\ &= \Sigma x^2-n\bar{x}^2.\end{aligned}$$

Hence, the variance may be calculated by

$$S^2 = \frac{1}{n}\Sigma x^2-\bar{x}^2,$$

or

$$S^2 = \frac{1}{n}\Sigma x^2-\frac{1}{n^2}(\Sigma x)^2.$$

When using a desk calculating machine the second of the above formulae is best because it avoids squaring the rounding-off error in $\bar{x}$.

Thus, the variance of the weights of the fifteen florins is

$$\begin{aligned} S^2 &= \tfrac{1}{15}(12.05^2+12.05^2+\ldots+12.93^2)-\tfrac{1}{225}(12.05+12.05+\ldots+12.93)^2 \\ &= 159.37843-159.2644 \\ &= 0.11403 \text{ g}^2 \end{aligned}$$

and the standard deviation S = 0.338 g (3 decimals).

Note that on an electronic computer, where the number of significant figures (or decimal places) is fixed, it is best to compute

$$S^2 = \frac{1}{n}\,\Sigma(x-\bar{x})^2$$

using as many decimal places as possible for $\bar{x}$.

The alternative formulae may give S^2 negative if the variance is small.

4. Change of origin and unit. The calculation of the variance and standard deviation can be considerably simplified by converting $x_1, x_2, \ldots, x_n$ to $X_1, X_2, \ldots, X_n$ as in §2.

Thus, the sum of the squares of deviations from the mean

$$\begin{aligned} nS^2 &= \Sigma(x-\bar{x})^2 \\ &= \Sigma\{(A+BX)-(A+B\bar{X})\}^2 \\ &= \Sigma\{BX-B\bar{X}\}^2 \\ &= B^2\Sigma(X-\bar{X})^2. \end{aligned}$$

Hence the variance $$S^2 = B^2\left\{\frac{\Sigma(X-\bar{X})^2}{n}\right\}$$

and the standard deviation

$$\begin{aligned} S &= B\sqrt{\left\{\frac{\Sigma(X-\bar{X}^2)}{n}\right\}} \\ &= B\sqrt{\left\{\frac{\Sigma X^2}{n}-\bar{X}^2\right\}}. \end{aligned}$$

Suppose, in the case of the fifteen 10p pieces the weights $X_1, X_2, \ldots, X_n$ are taken as

−12 3 4 5 6 7 8 11 13 17 18 20 22 30 33,

then $A = 12.60,\quad B = 0.01,\quad \Sigma X = 185,\quad \Sigma X^2 = 4119,$

$\bar{X} = 12.33$ (2 decimals), $\Sigma X^2/15 = 274.6$

and $$\begin{aligned} S &= 0.01\sqrt{122.5} \\ &= 0.111 \text{ g} \quad \text{(3 decimals).} \end{aligned}$$

Note that the value of A is not used in the calculation of S.

5. Dispersion or variability. Inspection of the separate weights of the fifteen 10p pieces shows that the *least* and *greatest* weights recorded are 12.48 g and 12.93 g respectively. That is to say, the *range* of the weights is 12.48 g to 12.93 g, a difference of 0.45 g. On the other hand the range of the weights of the florins is 12.05 g to 12.93 g, a difference of 0.88 g. This indicates that the *dispersion* or *variability* in the weights of the group of florins is approximately twice that of the group of 10p pieces. Alternatively, the weights of the florins may be said to be more widely *dispersed* or *spread* than the weights of the 10p pieces.

Some readers, no doubt, will be surprised that the weights of new 10p pieces vary so much. The weight of a new coin, like the diameter, is so often accepted as a *standard.* The mind often tends to accept the central tendency of a group as a rigid and unvarying standard. In Statistics we are particularly interested in the degree of departure from the central tendency. That is to say, we study and measure the dispersion or variability within the group. The universally accepted measures of dispersion are the variance and standard deviation defined in §3.

The range is unsatisfactory because it is based entirely on the two extreme members which may be abnormal. Moreover, the range generally depends on the size of the sample. If, in assessing the dispersion in a group, use is made of the range, it is generally done by first converting the range to the standard deviation by a table such as that given on page 112.

Measures of dispersion other than the range, variance and standard deviation were described fully in *A First Course in Statistics.* They will not be discussed further in this volume.

Let us note then that the standard deviation of the weights of the fifteen florins is 0.338 g, while the standard deviation of the weights of the fifteen 10p pieces is 0.111 g. This implies that the dispersion of the weights of the florins is approximately three times that of the 10p pieces. The full meaning of the standard deviation is not easily understood by the beginner. Only after studying a large number of examples such as those given in the first two chapters of this book will the student feel that some understanding of this measure of dispersion is being achieved.

The first four florins in the group of fifteen are obviously very worn. If these four observations are discarded, the mean and standard deviation of the other eleven are 12.815 g and 0.108 g. Thus the weights of the eleven florins in good condition have approximately the same location and dispersion as the weights of the fifteen new 10p pieces.

6. Exercises.

1. Working with 980 as origin and 0.1 as unit calculate the mean and the standard deviation of

980.8 981.1 980.7 980.3 981.8 982.5.

2. A sugar refiner uses machines which pack automatically 1 kilo cartons of sugar. To check that the machines are giving correct weight, cartons are selected at random and weighed accurately. The results of checking two machines are:

Accurate weights in kilos of eleven cartons

Machine *A*	1.017	1.051	1.078	0.996	1.033	1.059
Machine *B*	0.995	1.009	1.028	1.036	1.000	1.017
Machine *A*	1.082	1.014	1.040	1.072	0.998	
Machine *B*	1.027	1.045	1.006	1.018	1.039	

Calculate the means and the standard deviations and explain briefly what inference can be drawn from them.

3. A firm which manufactures lead-covered submarine cables checked the thickness of the cover on two of its cables by taking measurements in ten places:

Ten measurements, in cm of the thickness

Cable *A*	0.74	0.76	0.78	0.70	0.72
Cable *B*	0.70	0.72	0.73	0.74	0.72
Cable *A*	0.73	0.75	0.77	0.79	0.71
Cable *B*	0.72	0.74	0.71	0.72	0.73

Calculate for each cable the mean thickness of the cover and the standard deviation of the thickness. Explain briefly what deductions can be made about the covers of the two cables.

4. The times, in minutes, of a car journey made between 5.30 p.m. and 6.30 p.m. along the same route on five consecutive Mondays were as follows:

33 28 26 35 38.

Calculate the mean and the standard deviation of the times.

5. Find the mean and standard deviation of the set of numbers 8, 9, 10, 11, 12. From this set, ten samples each containing two numbers can be selected. Find the mean of each of these samples and calculate the standard deviation of these means. [London]

6. Two forms, one of 20 boys and the other of 30 boys, are given an examination. In the smaller form the average mark was 60 and the standard deviation was 7.0. In the other form the average mark was 50 and the standard deviation was 10.0. Find the standard deviation for the marks of the 50 boys taken as a single group. [London]

7. The numbers of members, means and standard deviations of three distributions are:

No. of members	280	350	630
Means	45	54	49
Standard deviations	6	4	8

Find the mean and standard deviation of the distribution formed by the three distributions taken together. [London]

8. The mean ages of n_1 boys is M_1, and the standard deviation from the mean of their age distribution is σ_1. The mean of the ages of n_2 girls is M_2, and the corresponding standard deviation from the mean is σ_2. Find the mean of the ages of the boys and girls combined.

If $M_1 = M_2$, obtain an expression for the standard deviation from the mean of the combined age distribution. [London]

9. The marks obtained by ten boys in an examination were 12, 17, 20, 23, 26, 29, 29, 35, 38, 41.

Find the standard deviation. The marks are now to be adjusted so that the mean is 60 and the standard deviation is 15. Calculate the highest and the lowest marks obtained on the new scale. What is the purpose of this adjustment? [London]

10. The table gives the number of minutes late or early for the arrival of a train on a number of runs:

Late	2	4	1	6	9	2	1	0
Early	3	1	.	.	.	.	.	.

Calculate the mean of these and the standard deviation.

After two more runs neither the mean nor the standard deviation is altered. Calculate, to the nearest half, the number of minutes late or early for each of these runs. [London]

11. Four boys sit for an examination. The average of their marks is M and the standard deviation is σ. The marks are converted to a new scale by the formula

$$y = 50-20(M-x)/\sigma,$$

where y is the new mark and x is the original mark. Find the mean and the standard deviation of the new marks.

If the original marks were 47, 57, 65, 71, find the new marks each to the nearest integer. [London]

12. The marks obtained by the n candidates who passed an examination but did not reach the credit standard ranged from 68 to 76. They were converted to a range of 50 to 60 by reading off the new mark, y, corresponding to an old mark, x, from the straight line graph joining the point (68, 50) to the point (76, 60). Find a formula for y in terms of x and deduce relationships between (i) the new mean, $\bar{y}$, and the old mean, $\bar{x}$; (ii) the new standard deviation, s', and the old standard deviation, s. [Northern]

13. Two dice, each of which has its faces numbered from 1 to 6, are thrown together, and the score found by squaring the difference between the numbers on the faces resting uppermost. Draw up a table showing the number of ways in which each possible score can be obtained. Determine the mean score and calculate the standard deviation from this mean. [London]

14. If the mean of two numbers x and y is 4, show that the mean of the four numbers x, $2x$, y and $2y$ is 6.

If the variance of the numbers x and y is 2, find the variance of the numbers x, $2x$, y and $2y$. [London]

15. Two numbers x and y have a mean 3 and variance 4. Three different numbers a, b and c have a mean 8 and a variance 36. Calculate the mean and variance of the five numbers x, y, a, b and c taken as one group. [London]

16. Given, as in §3, that $\bar{x}$ and S^2 are the mean and variance of the n observations

$$x_1, x_2, \ldots, x_n$$

and given also that the *mean-square deviation from an arbitrary value a* is

$$S_1^2 = \frac{1}{n}\Sigma(x-a)^2$$

show that

$$S^2 = S_1^2-(\bar{x}-a)^2.$$

7. From random sample to parent population. The fifteen 10p pieces and the fifteen florins are only of interest in that they enable us to make statements about 10p pieces and florins *in general.* They are *random samples* from which we are trying to estimate the properties of the *parent populations* from which they are drawn. To distinguish clearly between the mean and standard deviation of a random sample and the corresponding mean and standard deviation of its parent population English small italic letters are used for sample values and the corresponding Greek small letters for the population values. Thus m (rather than $\bar{x}$) and s are used to denote the mean and standard deviation of the sample while μ and σ are used to denote the mean and standard deviation of the parent population. The values μ and σ are examples of *parameters.* A parameter is a constant which takes different values for different populations.

Now m gives a good estimate of μ and we write

$$m = \frac{1}{n}\Sigma x = \text{Est}\ (\mu).$$

Thus 12.62 g may be accepted as an estimate of the mean weight of florins in general and 12.72 g as the mean weight of 10p pieces in general. At this stage the 12.723 g has been reduced to two decimal places because it seems reasonable to state the estimate with the same accuracy as the data.

For σ, however, a better estimate is obtained by using a divisor $(n-1)$ instead of n and denoting the value of the standard deviation obtained in this way by a small s. Thus

$$s = \sqrt{\left\{\frac{\Sigma(x-\bar{x})^2}{(n-1)}\right\}} = \text{Est}(\sigma),$$

or
$$s = \sqrt{\frac{n}{(n-1)}}\,S = \text{Est}(\sigma).$$

The reason for this is that σ should really be calculated by taking deviations from μ. However, we use m instead of μ and when the n deviations

$$(x_1-m),\quad (x_2-m), \ldots, (x_n-m)$$

are used their sum is zero. This implies that when $(n-1)$ of the deviations have been written down the nth deviation is predetermined and we say that, for a random sample of n observations, only $(n-1)$ *degrees of freedom* are available for the calculation of s. For example, in the case of the weights of the fifteen florins the deviations, in g, from the mean are: -0.57, -0.57, -0.54, -0.47, -0.02, $+0.08$, $+0.09$, $+0.10$, $+0.21$, $+0.24$, $+0.27$, $+0.28$, $+0.29$, $+0.30$, $+0.31$.

The sum of the first fourteen deviations is -0.31 and, therefore, the fifteenth deviation is predetermined as $+0.31$. The above explanation is not a proof but it is hoped that it will be sufficient to convince students at this stage of their course.

When calculating s, the formula

$$s = \sqrt{\left\{\frac{\Sigma(x-\bar{x})^2}{(n-1)}\right\}} = \text{Est}(\sigma)$$

is generally used in its equivalent form

$$s = \sqrt{\left\{\frac{\Sigma x^2}{(n-1)} - \frac{n}{(n-1)}\bar{x}^2\right\}} = \text{Est}(\sigma),$$

or
$$s = \sqrt{\left\{\frac{\Sigma x^2}{(n-1)} - \frac{(\Sigma x)^2}{n(n-1)}\right\}} = \text{Est}(\sigma).$$

The last form is convenient when using a desk calculating machine. To estimate the standard deviation of the weights of 10p pieces, therefore, the divisor 14 is used instead of 15 giving

$$\sigma = 0.115 \text{ g} \quad \text{(3 decimals)}.$$

Similarly, the estimate of the standard deviation of the weights of florins

$$\sigma = 0.350 \text{ g} \quad \text{(3 decimals)}.$$

8. Large samples. When n is large the estimate of σ obtained by using the divisor $(n-1)$ will differ very little from the estimate obtained by the divisor n.

Thus, when n is large,

$$\sqrt{\frac{n}{(n-1)}} = \left(1-\frac{1}{n}\right)^{-\frac{1}{2}}$$

$$= 1+\frac{1}{2n} \quad \text{approximately}$$

and the relation $$s = \sqrt{\frac{n}{(n-1)}}\,\mathrm{S} = \mathrm{Est}\,(\sigma) \quad \text{of §7}$$

becomes $$s = \left(1+\frac{1}{2n}\right)\mathrm{S} = \mathrm{Est}\,(\sigma).$$

Hence, when $n = 50$ the difference between s and S is 1 % and when $n = 100$ the difference is 0.5 %. Moreover, s is merely an estimate and so for *large samples* (say $n = 50$ or more) it is customary to use the divisor n.

9. Exercises.

1. The values given in §6, Ex. 1, are the results of six determinations of g by a particular piece of apparatus. Estimate from them the standard deviation, σ, to be expected if a large number of determinations of g were made using the same apparatus.

2. Use the times given in §6, Ex. 4, to estimate the standard deviation of the parent population. Explain briefly the meaning, in this example, of the phrase 'parent population'.

3. Estimate the mean and standard deviation of the whole output for each of the machines A and B of §6, Ex. 2.

4. In §6, Ex. 3, if it were possible to measure the thickness of the cover in a very large number of places throughout the whole length of each cable, what would you expect the mean and standard deviation, in each case, to be.

10. Frequency distribution with unequal group intervals. The results of a count of craters on the surface of the Moon are shown in table 1 A. In it the total number of 1596 craters have been divided into *groups* whose sizes are given in the left-hand column. Note that in the first of these groups, since the values are given to one decimal place, the range 5.0–9.9 implies that the diameters are greater than 4.95 and less than 9.95. Thus the range of the first group is 5. Note also that the *group intervals* in the left-hand column are unequal, the first being 5, the next seven being 10, the ninth being 20 and the tenth being 50. The numbers in the right-hand column indicate the *frequency* with which each particular size occurs. In table 1 A then, the 1596 separate observations have been grouped into a *frequency distribution.*

TABLE 1A

Frequency distribution of the diameters of the craters on the surface of the Moon

Diameter of crater (km)	Number of such craters (frequency)
5.0–9.9	650
10.0–19.9	438
20.0–29.9	223
30.0–39.9	105
40.0–49.9	58
50.0–59.9	28
60.0–69.9	30
70.0–79.9	21
80.0–99.9	23
100.0–149.9	20
Total	1596

11. Histogram. A convenient way of representing a frequency distribution diagrammatically is by a *histogram* as shown in fig. 1. In this type of diagram the frequencies 650, 438, ..., 23, 20 are represented by the *areas* of rectangles. Because the *widths* of the rectangles are 5, 10, ..., 20, 50 their respective *heights* are 130, 43.8, ..., 1.15, 0.4. Thus the sum of the areas of the 10 rectangles represents the total frequency 1596.

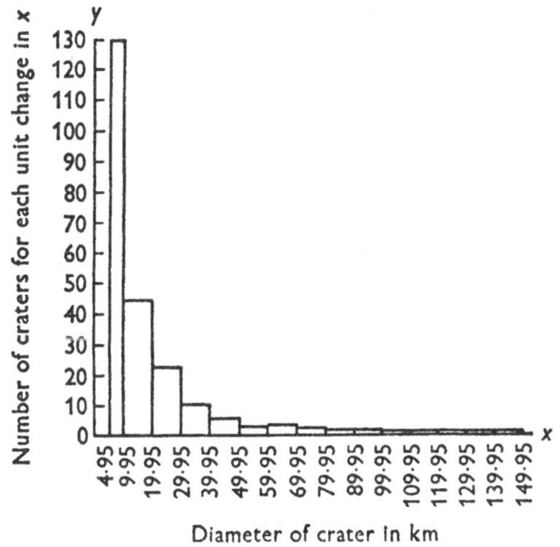

Fig. 1. Frequent distribution of the sizes of the craters on the surface of the Moon displayed as a histogram.

12. Frequency distribution with equal group intervals. The chief purpose of the example of § 10 was to make clear the method of diagrammatic representation of a frequency distribution by areas of rectangles of varying widths. Table 1 B is an example of a frequency distribution in which the group intervals are all equal. In this case the rectangles of the histogram, fig. 2, are of equal width. The *frequency polygon* is also shown in fig. 2. This is an alternative method of illustrating the frequency distribution. The frequency polygon is obtained by plotting each frequency against the mid-point of its range. It is not necessary to draw both the histogram and frequency polygon. The one or the other is sufficient to illustrate the data. Instead of the histogram in fig. 1, a frequency polygon could have been used to illustrate the data of table 1 A.

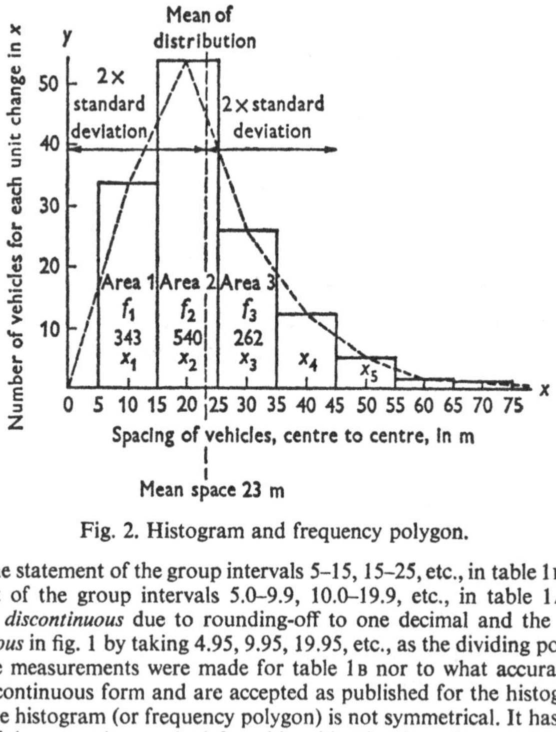

Fig. 2. Histogram and frequency polygon.

NOTE 1. The statement of the group intervals 5–15, 15–25, etc., in table 1 B differs from the statement of the group intervals 5.0–9.9, 10.0–19.9, etc., in table 1 A. The latter appear to be *discontinuous* due to rounding-off to one decimal and the histogram is made *continuous* in fig. 1 by taking 4.95, 9.95, 19.95, etc., as the dividing points. It is not clear how the measurements were made for table 1 B nor to what accuracy. They are presented in continuous form and are accepted as published for the histograms.

NOTE 2. The histogram (or frequency polygon) is not symmetrical. It has a longer tail to the right of the mean than to the left and is said to be *skewed positively* (or skewed to the right).

NOTE 3. Approximately 5 % of the observations lie outside the range *mean* ± 2 × *standard deviation* and, due to the skewing, all 5 % are above the range and none below. The student will be interested to refer back to this note when studying the normal distribution.

NOTE 4. The median is the value of x which bisects the total area of the histogram. By dividing the 15–25 group in the ratio 325:215 it can be estimated as 21 metres.

TABLE 1B

A traffic survey of the distances between vehicles travelling along a certain stretch of motorway

Spacing of vehicles, centre to centre, in metres	Number of such spaces (frequency)
5–15	343
15–25	540
25–35	262
35–45	122
45–55	51
55–65	11
65–75	7
Total	1336

13. Mean of frequency distribution. The definition of the mean stated in §1 can be extended to cover the case of a frequency distribution as follows:

IF THE OBSERVATIONS $x_1, x_2, \ldots, x_n$ HAVE RESPECTIVE FREQUENCIES $f_1, f_2, \ldots, f_n$ THE MEAN OF THE FREQUENCY DISTRIBUTION IS

$$\bar{x} = \frac{f_1x_1+f_2x_2+\ldots+f_nx_n}{f_1+f_2+\ldots+f_n}$$

$$= \frac{\Sigma fx}{\Sigma f}.$$

To apply this definition to the data of table 1B, $f_1, f_2, \ldots, f_n$ are taken as 343, 540, ..., 7 and $x_1, x_2, \ldots, x_n$ as 10, 20, ..., 70. That is to say x_1, x_2, etc., are the *mid-values* of the group intervals 5–15, 15–25, etc. This is illustrated diagrammatically in fig. 2. In making this assumption that each observation of a group takes the mid-value of the group errors are introduced. However, it was pointed out by W. F. Sheppard in 1898 that provided all the group intervals are equal the final error in the mean is negligible because the positive and negative errors in the individual observations tend to cancel each other.

14. First moments. By analogy with mechanics, since Σfx is the sum of the first moments of area of the histogram about the origin, $\bar{x} = \Sigma fx/\Sigma f$ is called the first moment of the distribution about the origin. The second, third and fourth moments will be defined later. Moments, of course, can be taken about points other than the origin and $\Sigma f(x-a)/\Sigma f$ is called the first moment of the distribution about $x = a$. An important and obvious property of first moments is that the first moment of any distribution about its mean is zero. In symbols, this is stated as $\Sigma f(x-\bar{x})/\Sigma f = 0$.

15. Variance and standard deviation of a frequency distribution. The definition of the variance stated in §3 can be extended to cover the case of the frequency distribution as follows:

IF THE OBSERVATIONS $x_1, x_2, \ldots, x_n$ HAVE RESPECTIVE FREQUENCIES $f_1, f_2, \ldots, f_n$ THE VARIANCE OF THE FREQUENCY DISTRIBUTION IS GIVEN BY

$$S^2 = \frac{f_1(x_1-\bar{x})^2+f_2(x_2-\bar{x})^2+\ldots+f_n(x_n-\bar{x})^2}{f_1+f_2+\ldots+f_n}$$

$$= \frac{\Sigma f(x-\bar{x})^2}{\Sigma f}.$$

Also since

$$\begin{aligned}\Sigma f(x-\bar{x})^2 &= \Sigma fx^2-2\bar{x}\Sigma fx+\bar{x}^2\Sigma f\\ &= \Sigma fx^2-2\bar{x}(\bar{x}\Sigma f)+\bar{x}^2\Sigma f\\ &= \Sigma fx^2-\bar{x}^2\Sigma f\end{aligned}$$

it can be seen that a useful alternative formula for the variance is

$$S^2 = \frac{\Sigma fx^2}{\Sigma f}-\bar{x}^2,$$

or

$$S^2 = \frac{\Sigma fx^2}{\Sigma f}-\left(\frac{\Sigma fx}{\Sigma f}\right)^2.$$

When this formula is applied to data such as that of table 1 B two important facts need stressing:

(i) Since Σf is large, the argument of §8

$$S = s = \text{Est}(\sigma).$$

(ii) Unlike §13, the errors introduced by the assumption that each observation takes the mid-value of the group are all made positive due to squaring. Consequently the final error in the variance is not negligible. For frequency distributions with equal group intervals an allowance can be made for the error thus caused by grouping. It is to *reduce the variance by* $\frac{1}{12}C^2$ *where C is the length of the group interval.* This is known as Sheppard's correction. The formula for the standard deviation thus becomes

$$S = \sqrt{\left\{\frac{\Sigma fx^2}{\Sigma f}-\left(\frac{\Sigma fx}{\Sigma f}\right)^2-\tfrac{1}{12}C^2\right\}}.$$

It will be realised that *working units* are often chosen so that $C = 1$. A full discussion of the correction is to be found in the article by W. F. Sheppard, 'On the calculation of the most probable values of the frequency constants from data arranged according to equidistant divisions of the scale'. *Proc. Lond. Math. Soc.* vol. 29, p. 353. It is also fully discussed by Professor A. C. Aitken in *Statistical Mathematics* (Oliver and Boyd).

16. Change of origin and unit. Table 1C shows the calculation of the mean and standard deviation of the frequency distribution given in table 1B. It will be seen that the arithmetic of the calculation has been reduced by changing the origin and unit.

If $A = 40$ metres and $B = 10$ metres are the arbitrary origin and unit respectively, then by §2

$$\begin{aligned}\bar{x} &= A+B\bar{X}\\ &= 40+10\left\{\frac{\Sigma fX}{\Sigma f}\right\} \text{ metres}\\ &= 40-\frac{10\times 2277}{1336} \text{ metres}\\ &= 22.9 \text{ metres} = \text{Est}(\mu).\end{aligned}$$

TABLE 1C

Calculation of the mean and the standard deviation

Spacing of vehicles centre to centre in metres (mid-values) x	Number of such spaces (frequency) f	Spacing with 40 metres as origin, 10 metres as unit X	First moment of each value of x fX	Second moment of each value of x fX^2
10	343	−3	−1029	3087
20	540	−2	−1080	2160
30	262	−1	−262	262
40	122	0	0	0
50	51	1	51	51
60	11	2	22	44
70	7	3	21	63
Total	$\Sigma f = 1336$		$\Sigma fX = -2277$	$\Sigma fX^2 = 5667$

Also by §§4 and 15,

$$\begin{aligned}\mathrm{S} &= B\sqrt{\left\{\frac{\Sigma fX^2}{\Sigma f}-\bar{X}^2-\frac{c^2}{12}\right\}}\\ &= 10\sqrt{\left\{\frac{5667}{1336}-\left(-\frac{2277}{1336}\right)^2-\frac{1}{12}\right\}}\end{aligned}$$

since, for the X values, the group-interval $c = 1$. Thus

$$\mathrm{S} = 11.2 \text{ metres} = \text{Est}(\sigma).$$

17. Second, third and fourth moments. Continuing the analogy with mechanics, $f_1x_1^2, f_2x_2^2, \ldots, f_nx_n^2$ are the second moments of the areas $f_1, f_2, \ldots, f_n$ about the origin and $\Sigma fx^2/\Sigma f$ is called the second moment of the distribution about the origin. The second moment of the distribution about $x = a$ is $\{\Sigma f(x-a)^2/\Sigma f\}$. It will be realised that the second moment of the distribution about the mean, $\{\Sigma f(x-\bar{x})^2\}/\Sigma f$, is the variance. It is customary to represent the first and second moments about the mean by the symbols μ_1 and μ_2 and to use μ_1' and μ_2' (or ν_1 and ν_2) to represent the corresponding moments about other values.

So far, then, it has been established that for all distributions μ_1 is zero and μ_2 is the variance. The third and fourth moments about the mean, μ_3 and μ_4, are defined by

$$\mu_3 = \frac{\Sigma f(x-\bar{x})^3}{\Sigma f},$$

$$\mu_4 = \frac{\Sigma f(x-\bar{x})^4}{\Sigma f}.$$

Their calculation is illustrated by §18, Ex. 11.

It may seem strange that the symbol μ_1 has been introduced for a *statistic* which is known to be zero. It does, however, complete the set of the four moments about the mean

$$\mu_1, \quad \mu_2, \quad \mu_3, \quad \mu_4$$

corresponding to the four moments about the origin (or about any other convenient value)

$$\nu_1, \quad \nu_2, \quad \nu_3, \quad \nu_4.$$

18. Exercises.

1. The following table gives the distribution of the total population (in thousands) of Northern Ireland 30 June 1958:

Age	0–4	5–14	15–19	20–34	35–49	50–64	65–74	75–84
Number	140	265	116	276	258	209	85	45

Construct a histogram to illustrate the distribution. [London]

2. The following table gives the distribution of the marks of 942 candidates in an examination:

Marks	No. of candidates
90–99	2
80–89	10
70–79	51
60–69	125
50–59	179
40–49	227
30–39	181
20–29	108
10–19	45
0–9	14

Calculate the mean and the standard deviation. [London]

3. The following table gives the distribution by seating capacity of new bus and coach registrations during 1959:

Seating capacity	No. of vehicles
15–20	3
21–26	2
27–32	69
33–40	390
41–48	1823
49–56	25

Illustrate the data by a frequency polygon and calculate the mean and the standard deviation of the distribution. [London]

4. The following table gives the distribution of the ages of persons in Northern Ireland under the age of 35 on 30 June 1960:

Age	Number (in thousands)
0–4	144
5–9	132
10–14	133
15–19	123
20–24	103
25–29	88
30–34	86

Calculate the mean and the standard deviation of this distribution. [London]

5. The following table gives the 1964 estimates of the age distribution of the home population of the United Kingdom. (Source: *Annual Abstract of Statistics*, 1966.)

Age group	Number (thousands)	Age group	Number (thousands)
under 1	976	40–44	3857
1 and under 2	960	45–49	3250
2–4	2713	50–54	3616
5–9	4020	55–59	3424
10–14	3848	60–64	2996
15–19	4257	65–69	2327
20–24	3529	70–74	1788
25–29	3382	75–79	1241
30–34	3346	80–84	699
35–39	3459	85 and over	377

Draw a histogram and comment on any special features of the diagram. (You may assume that '85 and over' means '85–100'.)

6. The following frequency distribution was obtained from the results of compressive strength tests on 100 cubes of concrete:

Compressive strength (kilos per sq cm)	Number of cubes
160–179	1
180–199	7
200–219	26
220–239	38
240–259	22
260–279	6

Calculate the mean compressive strength and the standard deviation of the compressive strengths. By means of a histogram, or otherwise, show that about half of the cubes lie within the limits

mean ± two-thirds of the standard deviation. [A.E.B.]

7. The following frequency table was obtained in a test of 'distances to take-off' of a hundred aircraft of the same type:

Distance to take-off in metres	Number of aircraft
320–330	2
330–340	4
340–350	9
350–360	14
360–370	19
370–380	19
380–390	15
390–400	10
400–410	5
410–420	2
420–430	1

Calculate the mean and the standard deviation of the distances. [Northern]

8. Measurements are made to the nearest cm of the heights of 100 children. Draw the frequency diagram of the following distribution:

Height	160	161	162	163	164	165	166	167	168
Frequency	2	0	15	29	25	12	10	4	3

Calculate the mean, and the standard deviation from the mean. [London]

9. The following values of a quantity x were obtained experimentally:

x	18	19	20	21	22	23	24	25	26
Frequency	1	5	8	12	10	7	4	1	2

Calculate the mean and the standard deviation and draw the histogram for this distribution. [London]

10. The numbers 1, 2, 3, 4, 5, 6 are printed one on each of six cards. Three cards are drawn at random and the numbers on each of them are added together. In 100 trials the following distribution is obtained:

Total	6	7	8	9	10	11	12	13	14	15
Frequency	3	6	9	16	15	11	17	12	5	6

Calculate the mean and standard deviation of this distribution. Would you expect the mean to rise or fall if the number of trials were greatly increased, assuming that every selection is equally probable? [London]

11. Given, as in §§15, 18:

that (i) the observations $x_1, x_2, \ldots, x_n$ have respective frequencies $f_1, f_2, \ldots, f_n$;

that (ii) the first, second, third and fourth moments about the mean $\bar{x}$ are

$$\mu_1 = \frac{\Sigma f(x-\bar{x})}{\Sigma f}, \qquad \mu_2 = \frac{\Sigma f(x-\bar{x})^2}{\Sigma f},$$

$$\mu_3 = \frac{\Sigma f(x-\bar{x})^3}{\Sigma f}, \qquad \mu_4 = \frac{\Sigma f(x-\bar{x})^4}{\Sigma f},$$

and that (iii) the first, second, third and fourth moments about the origin are respectively

$$\nu_1 = \frac{\Sigma fx}{\Sigma f}, \qquad \nu_2 = \frac{\Sigma fx^2}{\Sigma f}, \qquad \nu_3 = \frac{\Sigma fx^3}{\Sigma f}, \qquad \nu_4 = \frac{\Sigma fx^4}{\Sigma f},$$

prove that

$$\mu_1 = 0,$$

$$\mu_2 = \nu_2 - \nu_1^2,$$

$$\mu_3 = \nu_3 - 3\nu_1\nu_2 + 2\nu_1^3,$$

$$\mu_4 = \nu_4 - 4\nu_1\nu_3 + 6\nu_1^2\nu_2 - 3\nu_1^4.$$

12. The table shows the intelligence quotient (I.Q.) of 100 pupils at a certain school. Calculate (i) the mean, (ii) the mean deviation,* (iii) the standard deviation.

I.Q.	55–	65–	75–	85–	95–	105–	115–	125–	135–
No. of pupils	1	3	7	20	32	25	10	1	1

NOTE: 55– means 'from 55.0 to 64.9 inclusive', each I.Q. being given correct to one decimal place. [Cambridge]

13. The following distribution of marks (out of 100) was obtained with a certain examination paper:

Mark range	10–29	30–39	40–49	50–59	60–69	70–79	80–89
Frequency	2	8	14	26	28	10	4

Present these marks in a histogram.

Calculate the mean mark as well as you can, explaining the limitations of your calculation. Do you consider that the mean or the median is the better measure of average performance for these observations? Give reasons for your answer. [Cambridge]

* See the Glossary for the definition.

14. If the frequency of a measurement x is f and x_0 is any number, prove that the mean value M of the measurements is equal to

$$N^{-1}\Sigma f(x-x_0)+x_0,$$

where $N = \Sigma f$ and the summations include all non-zero values of f. Prove also that the standard deviation σ of the measurements is given by

$$\sigma^2 = N^{-1}\Sigma f(x-x_0)^2-(x_0-M)^2.$$

The frequency f with which x α-particles are emitted from a radioactive specimen in a given time is shown in the following table:

x	0	1	2	3	4	5	6	7	8	9	10	11	12
f	57	203	383	525	532	408	273	139	45	27	10	4	2

Using $x_0 = 4$ as a trial mean, or otherwise, calculate the mean and standard deviation of the observed values of x. [Cambridge]

15. If the 'crude' mark x of each of the n pupils in a class is replaced by a 'standardised' mark $(x-m)/\sigma$, where m is the mean mark and σ the standard deviation of the marks in that particular subject, calculate (a) the mean, (b) the standard deviation, of the standardised marks.

By standardising the crude marks given in the following table, place the four boys A, B, C and D in an order of merit for tests in the three subjects, English, Mathematics and French, taken in the same examination.

	m	σ	A	B	C	D
English	55	10	60	50	60	70
Mathematics	70	4	80	60	70	80
French	60	12	50	80	60	40

[Cambridge]

2

THE NORMAL DISTRIBUTION

19. The normal probability curve. The equation of the normal probability curve in its most convenient form is

$$y = \frac{1}{\sqrt{(2\pi)}} e^{-\frac{1}{2}x^2}.$$

It was originally derived by Gauss as the law of the probable distribution of errors of measurement and hence it is often called the *Gaussian* or *error* curve. It is a continuous function for which x may range from $-\infty$ to $+\infty$, but the part of the curve which has real practical value lies between $x = -4$ and $x = +4$. Ordinary books of mathematical tables contain values of e^x and e^{-x} and the student will have no difficulty in using them to obtain values of the ordinate y of the normal probability curve for given values of x. The normal probability curve, however, is of such fundamental importance in statistical analysis that it is useful to have values of y, tabulated for given values of x as in table A1,* page 195.

20. Negative values of *x*. Since y is a function of x^2, negative values of x such as -1, -2, -3 give the same positive value to y as the positive values 1, 2, 3. Thus the graph of the function is symmetrical about the y-axis.

21. The graph of the function. The graph of

$$y = \frac{1}{\sqrt{(2\pi)}} e^{-\frac{1}{2}x^2}$$

between $x = -4$ and $x = +4$ is shown in fig. 3. The student can construct it for himself by using the values given in table A1.

22. The area under the curve. The dotted rectangles shown in fig. 3 are each of unit width and their heights are the ordinates at -3.5, -2.5, -1.5, -0.5, $+0.5$, $+1.5$, $+2.5$ and $+3.5$. Their areas are, therefore,

0.0009, 0.0175, 0.1295, 0.3521, 0.3521, 0.1295, 0.0175, 0.0009.

* Those table numbers preceded by a capital 'A' are to be found in the Appendix.

Thus the area under the graph between $x = -4$ and $x = +4$ is approximately unity. Note that the two rectangles between $x = -4$ and $x = -3$ and between $x = 3$ and 4 are so small that they have been left to the imagination.

Actually the area between $-\infty$ and $+\infty$ can be proved to be exactly unity. Mathematically this is stated as

$$\frac{1}{\sqrt{(2\pi)}}\int_{-\infty}^{\infty} e^{-\frac{1}{2}x^2}\,dx = 1.$$

The above approximation of unity for the area between -4 and $+4$ is a basis for the statement in §1 that the part of the curve which is of real practical value is that between -4 and $+4$.

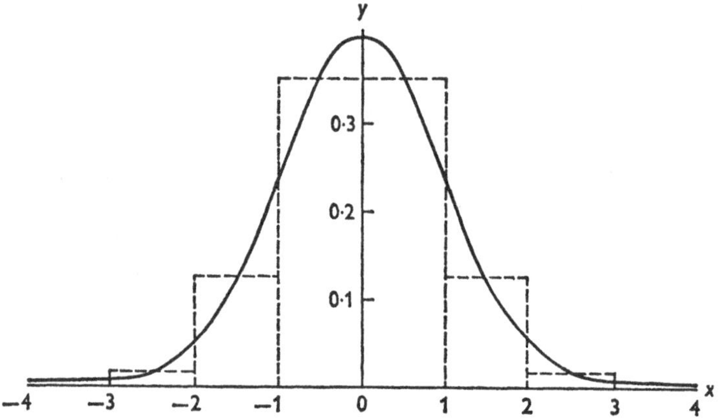

Fig. 3. The normal probability curve. The continuous bell-shaped curve is the graph of

$$y = \frac{1}{\sqrt{(2\pi)}}\,e^{-\frac{1}{2}x^2}.$$

The dotted rectangles are used to show that the area under the curve is approximately unity. The rectangle between $x = 3$ and $x = 4$ is left to the imagination and similarly that between $x = -4$ and $x = -3$.

23. Exercises using Simpson's rule. Simpson's rule, that an approximation for the area A under $y = f(x)$ between two ordinates y_1 and y_3, at a distance $2h$ apart is

$$A = \tfrac{1}{3}h\{y_1 + 4y_2 + y_3\},$$

where y_2 is the mid-ordinate can be used, in conjunction with table A1, to obtain approximations for the area under the probability curve between given values of x. Thus, for the area between $x = 0$ and $x = 1$,

$$2h = 1, \quad y_1 = 0.3989, \quad y_2 = 0.3521, \quad y_3 = 0.2420,$$

and an approximation for the area is

$$A = \tfrac{1}{6}\{0.3989 + 1.4084 + 0.2420\}$$
$$= 0.3416.$$

The student will find it a useful exercise to verify, by similar calculations, that

(i) the area between $x = 1$ and $x = 2$ is approximately 0.1357,

(ii) the area between $x = 2$ and $x = 3$ is approximately 0.0214,

(iii) the area between $x = 3$ and $x = 4$ is approximately 0.00135.

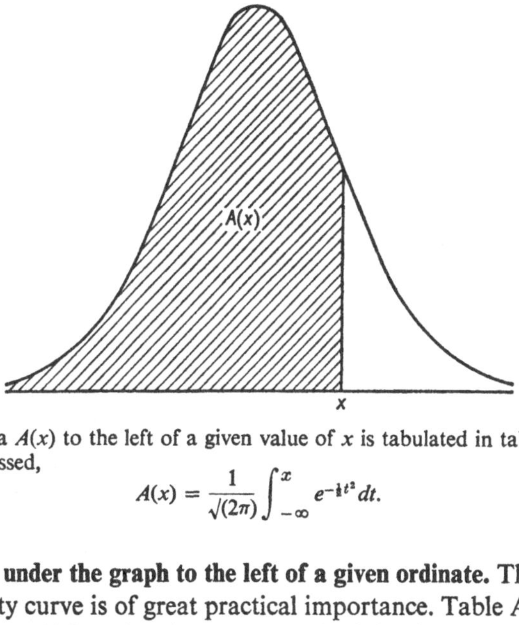

Fig. 4. The area $A(x)$ to the left of a given value of x is tabulated in table A2. Mathematically expressed,

$$A(x) = \frac{1}{\sqrt{(2\pi)}}\int_{-\infty}^{x} e^{-\frac{1}{2}t^2}\,dt.$$

24. The area under the graph to the left of a given ordinate. The area under the probability curve is of great practical importance. Table A2, page 196, shows the area $A(x)$ under the graph to the left of a given value of x as illustrated in fig. 4. Thus $A(0) = 0.5000$ indicates that half of the area under the graph is to the left of $x = 0$ and $A(1) = 0.8413$ indicates that 0.8413 of the area (i.e. 84.13 %) is to the left of $x = 1$. Moreover, $A(1) - A(0) = 0.3413$ indicates that 0.3413 of the area (i.e. 34.13 %) is between the ordinates at $x = 0$ and $x = 1$.

25. $A(x)$ for negative value of x. Table A2 indicates that $A(1.96) = 0.975$. Thus 97.5 % of the area is to the left of $x = 1.96$ and 2.5 % to the right of $x = 1.96$. As the graph is symmetrical about the y-axis it follows that 2.5 % of the area lies to the left of $x = -1.96$ and hence

$$A(-1.96) = 1 - A(1.96).$$

In general it may be stated that the area to the left of a negative value of x is obtained by subtracting from unity the area to the left of the corresponding positive value of x. In symbols this is written

$$A(-x) = 1 - A(x)$$

and it is illustrated diagrammatically in fig. 5.

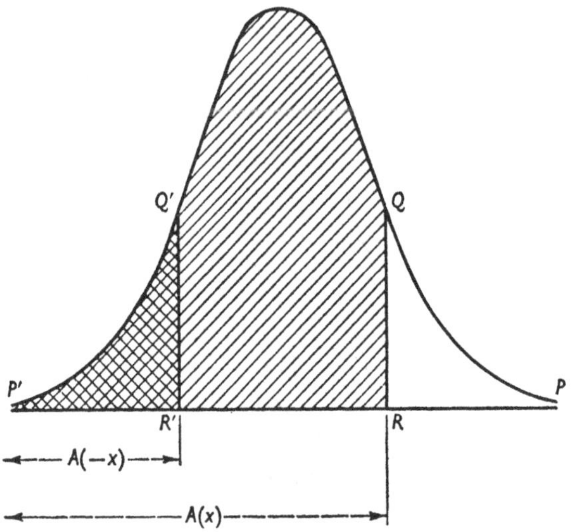

Fig. 5. Illustrates that $A(-x) = 1 - A(x)$ because area PQR = area $P'Q'R'$.

26. The mean and the standard deviation. The normal probability distribution given in tables A1 and A2 has its mean situated at the origin, $x = 0$, and its standard deviation is unity. If the heights of a group of men are known to be normally distributed about a mean of 168 cm with a standard deviation of 5 cm then heights in inches of 168, 168+5, 168+10, 168+15, 168+20 must be regarded as replacing $x = 0, 1, 2, 3, 4$ and

$$A(1) - A(0) = 0.3413$$

indicates that 0.3413 of the total number of men in the group will probably be between 168 and 173 cm in height while

$$A(2) - A(1) = 0.1359$$

indicates that 0.1359 of the total number will probably be between 173 and 178 cm in height and so on.

The examples and exercises in the following paragraphs illustrate various practical applications of the mean and standard deviation of a normal distribution.

27. How to construct a normal frequency distribution when given its mean and its standard deviation. *Jackets for young men are made in the following sizes, according to chest measurement*

Size	1	2	3	4	5	6
Chest measurement (cm)	75–	80–	85–	90–	95–	100–105

The distribution of chest measurements of young men in a certain age range is known to be approximately normal. Assuming a normal distribution with mean 89.075 *cm and standard deviation* 5.00 *cm, estimate the percentages of young men in this age range likely to require each of the six sizes and also the percentages likely to fall above or below the size range.* [Northern]

TABLE 2A

Chest measurement x (cm)	Deviation of chest measurement from mean $x-\bar{x}$ (cm)	Standardised chest measurement $\frac{x-\bar{x}}{\sigma} = X$	Value from table A2 of $A(X)$	$A(X)$ expressed as a percentage $100A(X)$
75	−14.075	−2.815	0.00244	0.244
80	− 9.075	−1.815	0.0347	3.47
85	− 4.075	−0.815	0.2075	20.75
90	+ 0.925	+0.185	0.5734	57.34
95	+ 5.925	+1.185	0.8820	88.20
100	+10.925	+2.185	0.98556	98.556
105	+15.925	+3.185	0.99927	99.927

The first step of the calculation is seen in table 2A. The chest measurements x are set down in column one with their deviations from the mean chest measurement $\bar{x}$ in column two. These deviations are then *standardised* by dividing them by the standard deviation σ. The standardised chest measurements X of the third column are then used to obtain from table A2 the values $A(X)$ of column four. In column five the values $A(X)$ are finally expressed as percentages. The percentage of young men likely to require size 1 is then

$$100A(-1.815)-100A(-2.815) = 3.23$$

and the percentage likely to require size 2 is

$$100A(-0.815)-100A(-1.815) = 17.28.$$

By continuing this process the percentages of young men likely to require each of the six sizes can be obtained and set out as shown in table 2B. Also, the percentage likely to fall below the size range is

$$100A(-2.185) = 0.244,$$

while the percentage likely to fall above the size range is

$$100 - 100A(3.185) = 0.073.$$

These figures are also included in table 2B.

28. The histogram. Table 2B can be illustrated diagrammatically as the *histogram* shown in fig. 6. Note that, although the distribution is normal, the histogram is not symmetrical because the size divisions are not symmetrical about the mean.

TABLE 2B

Chest measurement (cm)	Below 75	75–	80–	85–	90–	95–	100–105	Above 105
Size	—	1	2	3	4	5	6	—
Percentage of young men requiring the size	0.24	3.23	17.28	36.59	30.86	10.37	1.37	0.07

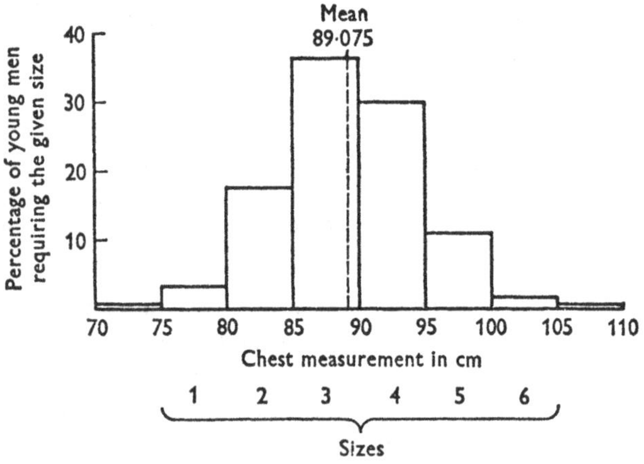

Fig. 6. Histogram showing the distribution in sizes of jackets for young men.

29. Exercises.

1. A sugar refiner uses a machine which packs automatically 1 kilo cartons of sugar. When the machine is in operation the cartons emerge from it in a continuous stream ready for sale. In order to check that the machine is giving correct weight sample cartons are taken at random from it and weighed accurately. From these random samples it is estimated that the mean weight of carton delivered by the machine is 1.02 kilos and that the standard deviation of the weights of the cartons is 0.008 kilos. Assuming that the weights are normally distributed construct a frequency table showing the percentage of cartons the machine delivers (i) less than 0.99 kilos, (ii) between 0.99 kilos and 1.00 kilos, (iii) between

1.00 kilos and 1.01 kilos, (iv) between 1.01 kilos and 1.02 kilos, (v) between 1.02 kilos and 1.03 kilos, (vi) between 1.03 kilos and 1.04 kilos, (vii) between 1.04 kilos and 1.05 kilos, (viii) over 1.05 kilos.

Illustrate diagrammatically by drawing a histogram.

2. The mean *length of life* of a certain type of television tube is 1600 h with a standard deviation of 250 h. Assuming that the lengths of life of television tubes of this type are normally distributed, copy, and complete, the following frequency table and use it to draw a histogram:

Length of life (h)	Less than 750	750– 1000	1000– 1250	1250– 1500	1500– 1750	1750– 2000	2000– 2250	More than 2250
Percentage of tubes								

3. The *recovery time* of an aircraft is the time between landing and being ready to fly again. In an investigation on a certain type of aircraft it was found that the recovery times were normally distributed about a mean of 19 min with a standard deviation of 3 min. Make out a frequency table showing the percentage of aircraft with recovery times 5 min to 10 min, 10 min to 15 min, 15 min to 20 min, etc., and draw the histogram.

30. An example in which it is assumed that the variability of production can be controlled. *The heights in inches of 'one inch' compression springs produced in a workshop may be assumed as normally distributed about a mean length of* 1.005 *in. with standard deviation* 0.003 *in. Estimate the limits which will contain the central* 50 % *of production.*

If the mean remains unchanged, find the value to which the standard deviation must be reduced in order that these limits should contain 90 % *of production.*

The central 50 % of the normal probability curve lies between the values x_1 and x_2 where $A(x_1) = 0.25$ and $A(x_2) = 0.75$. Table A2 shows that $A(x_2) = 0.75$ when $x_2 = 0.675$. Thus $x_2 = 0.675$ and $x_2 = -0.675$ and the central 50 % of production lies between

$$(\text{mean} - 0.675\sigma) \quad \text{and} \quad (\text{mean} + 0.675\sigma),$$

where σ is the standard deviation.

The required range is, therefore,

$$(1.005 - 0.675 \times 0.003) \quad \text{to} \quad (1.005 + 0.675 \times 0.003),$$

i.e. 1.003 in. to 1.007 in. (to the nearest thousandth of an inch).

The central 90 % of the normal probability curve lies between the values x_1 and x_2 where $A(x_1) = 0.05$ and $A(x_2) = 0.95$. Table A2 shows that $A(x_2) = 0.95$ when $x_2 = 1.645$. Thus $x_2 = 1.645$ and $x_1 = -1.645$ and the central 90 % of production lies between

$$(\text{mean} - 1.645\sigma_1) \quad \text{and} \quad (\text{mean} + 1.645\sigma_1),$$

where σ_1 is the new standard deviation.

If these limits are the same as the previous limits

$$1.645\sigma_1 = 0.675\sigma$$

and

$$\sigma_1 = \frac{0.675\sigma}{1.645}$$

$$= 0.00123 \text{ in.}$$

A final summary of this example is as follows: *While the standard deviation remains at* 0.003 *in. only* 50 % *of the production lies between* 1.003 *in. and* 1.007 *in. but if the standard deviation can be reduced to* 0.001 *in. the variability of production will have been controlled to such an extent that more than* 90 % *of the production will lie between these limits.*

31. An example in which the mean is adjusted, the variability remaining constant. *Estimate the percentage of underweight packets which the machine described in* §29, *Ex.* 1, *is delivering.*

The Board of Trade will allow machines of this kind to deliver up to 5 % *of their packets underweight. If the machine is to comply with this regulation calculate the lowest value to which the mean may legally be lowered and estimate the amount of sugar that would thus be saved in* 1000 *packets.*

The specified weight of 1 kg is 0.02 kg *below* the mean and the standard deviation is 0.008 kg. The standardised value of 1 kg is, therefore, $-0.02/0.008 = -2.5$.

Now

$$A(-2.5) = 1 - A(2.5)$$

$$= 1 - 0.99379$$

$$= 0.00621.$$

Hence about 0.6 % of the packets are underweight.

If 5 % of the packets are to be underweight $A(x) = 0.05$ gives the standardized value x of 1 kg *with respect to the new mean.*

Therefore

$$A(-x) = 0.95$$

$$-x = 1.645$$

$$x = -1.645.$$

This indicates that 1 kg is 1.645×0.008 kg *below* the new mean and hence the new mean is 1.013 kg correct to 3 decimal places. When the machine is delivering packets of mean weight 1.02 kg, the amount of sugar given away as overweight in 1000 packets is

$$0.02 \times 1000 \text{ kg} = 20 \text{ kg}.$$

When the mean weight is reduced to 1.013 kg the amount given away is 13 kg. The adjustment of the mean thus effects a saving or 7 kg of sugar in 1000 packets.

32. Exercises.

1. Tests on electric lamps of a certain type indicated that their lengths of life could be assumed to be normally distributed about a mean of 1860 h with a standard deviation of 68 h. Estimate the percentage of lamps which can be expected to burn (i) more than 2000 h, (ii) less than 1750 h.

2. Steel rods which are being manufactured to a specification of 30 mm diameter are acceptable if they are within the *tolerance limits* 30.1 mm and 29.9 mm. If the diameters can be assumed to be normally distributed about a mean 30.02 mm with standard deviation 0.06 mm, estimate the percentage of rods that will be rejected (i) oversize, (ii) undersize.

3. Limit gauges are used to reject all components in which a certain dimension is greater than 60.4 mm or less than 59.6 mm. It is found that about 5 % are rejected oversize and 5 % are rejected undersize. Assuming that the dimensions are normally distributed, find the mean and standard deviation of the distribution.

Estimate what the percentages of rejects would be if the limits were (i) 60.3 mm to 59.7 mm, (ii) 60.5 mm to 59.5 mm.

4. A machine makes electrical resistors having a mean resistance of 100 ohms with a standard deviation of 5 ohms. It is observed that a certain firm rejects approximately 11 % of the resistors ($5\frac{1}{2}$ % over and $5\frac{1}{2}$ % under) because they are not within its *tolerance limits*. Assuming the distribution of values to be normal, estimate what tolerance limits are employed by the firm.

5. In trials of the effectiveness of detergents in washing-up the measure x of performance is the number of plates which can be washed before the foam on the water is reduced to a thin surface layer. For a certain detergent x is known to be approximately normally distributed with mean 25 and standard deviation 4. Draw the frequency curve for this distribution, taking a scale of 1 in. to 4 plates and making the area equal to 5 sq in.

In fifty trials of a new detergent x has the frequency distribution given below:

x	21–24	25–28	29–32	33–36	37–40
Frequency	1	4	37	6	2

On the same diagram as before draw a histogram of area 5 sq in. to represent these data. Describe in words the changes in performance resulting from the use of the new detergent. [Northern]

6. The weights of loaves of bread made in a large bakery can be assumed to be normally distributed with standard deviation 7 g, and the mean of the distribution is so placed that only 0.1 % of the loaves produced have weights falling below the statutory minimum of 800 g. The weekly output of loaves averages 250000 and the cost in £ of producing a loaf of weight w g is given by the formula $0.013+0.000017w$. If by the installation of new dough-dividing machinery the standard deviation could be reduced to 3 g, and the mean allowed to fall far enough to give the same percentage below the statutory minimum as before, determine the average saving in £ per week.

[Northern]

3

PROBABILITY

33. Casting a die. When an ordinary die is cast, each of the numbers *one*, *two*, *three*, *four*, *five*, *six* has an equal chance of falling uppermost. These six outcomes of the *trial* are said to be *exhaustive* and *mutually exclusive* because one of the six must occur and only one can occur. There is, therefore, 1 *chance out of* 6 of obtaining any particular number and we say that

$$\textit{the probability of throwing a six} = \tfrac{1}{6}.$$

This is often abbreviated to

$$P(\text{six}) = \tfrac{1}{6} \quad \text{or} \quad P(6) = \tfrac{1}{6}.$$

34. Playing cards. An ordinary pack of 52 playing cards contains 4 *aces* and hence, if we select at random 1 card from a well-shuffled pack, there are 4 chances out of 52 of it being an ace. Thus

$$\textit{the probability of selecting an ace} = \tfrac{4}{52} = \tfrac{1}{13},$$

or

$$P(\text{any ace}) = \tfrac{1}{13}.$$

Also,

$$\textit{the probability of selecting the ace of spades} = \tfrac{1}{52},$$

$$\text{or} \quad P(\text{ace of spades}) = \tfrac{1}{52}.$$

35. Tossing a coin. When a coin is tossed it may fall either as a *head* or a *tail* and thus

$$\textit{the probability of a head} = \tfrac{1}{2},$$

$$\text{or} \quad P(H) = \tfrac{1}{2}.$$

36. The meaning of probability. When we say that the probability of throwing a *six* is $\frac{1}{6}$ we do not mean that exactly 1 throw out of every 6 will produce a *six*. If the student actually throws a die 60 times it is *likely* that he will obtain approximately 10 *sixes*. If he throws one 600 times a *reasonable estimate* of the number of *sixes* he can expect is 100 but it is very unlikely that it will be exactly 100. The following table gives the results of an actual experiment in which a die was cast 648 times. The frequency of each number approximates to 108 which is what we should expect since the probability of each number is $\frac{1}{6}$:

Number thrown	1	2	3	4	5	6
Frequency	96	98	117	130	107	100

37. Definition of probability. Sections 33–35 are simple illustrations of *random processes*. A random process or *random experiment* is a repetitive process or operation that, in a single trial, may result in any one of a number of possible outcomes each determined by chance. Probability may be defined as follows:

If a random process can result in n equally likely and mutually exclusive outcomes and if a of these outcomes have an attribute A, then the probability of A is the ratio a/n and we write

$$P(A) = a/n.$$

Note that this definition assumes that n is finite and that we have some way of deciding whether or not the outcomes are equally likely. In the case of the playing cards we intuitively accept that the 52 outcomes are equally likely and that 4 have the attribute of being aces.

An alternative definition is based on continued experimentation and does not depend on the assumption of the outcomes being equally likely. Suppose we conduct a sequence of experiments such that:

expt. 1 consists of n_1 trials in which attribute A occurs a_1 times,

expt. 2 consists of n_2 trials in which attribute A occurs a_2 times and so on to

expt. k in which attribute A occurs a_k times in n_k trials.

The *separate relative frequencies* of A in the experiments are then

$$a_1/n_1,\quad a_2/n_2,\quad \ldots,\quad a_k/n_k$$

and the *combined* relative frequency at the end of the kth experiment is

$$f = \frac{a_1+a_2+\ldots+a_k}{n_1+n_2+\ldots+n_k} = \frac{\Sigma a}{\Sigma n}.$$

Although the separate relative frequencies may differ considerably the combined relative frequency tends to converge to some value, $P(A)$, which is the probability of A. For any finite sequence of trials the observed value of f is an estimate of $P(A)$. If the number of trials is large, f will be close to $P(A)$.

An illustration of this definition can be drawn from the table of 'Random Sampling Numbers' given in the *Cambridge Elementary Statistical Tables*, page 12. Note first the statement at the foot of the page that each digit is an independent sample from a population in which the digits 0 to 9 are equally likely; that is, each has a probability of 1/10. The truth of this statement need not be accepted until it has been demonstrated by the following experiment:

Read from left to right in the usual way and note that the digit 0 appears once in the first 4 digits; it does not appear at all in the next 8 digits; it appears twice in the next 12 digits and so on. Summarising for the first 1200 digits we get

Number of digits	4	8	12	16	40	120	200	200	600
Number of times 0 appears	1	0	2	1	6	9	15	31	53
Separate relative frequencies (2 decimals)	0.25	0	0.17	0.06	0.15	0.08	0.08	0.16	0.09

These results illustrate the fact that, although the separate relative frequencies differ considerably, the combined relative frequency after 1200 digits is 118/1200 which is close to 1/10.

38. Success and failure. If we regard the throwing of a *six* as a *success* and the throwing of any other number as a *failure*, the probability of success is 1/6 and the probability of failure is 5/6. It is customary to use the letter p for the probability of success and the letter q for the probability of failure. Thus $p = \frac{1}{6}$ and $q = \frac{5}{6}$ and $q = 1-p$.

Alternatively, we write $P(S)$ as the probability of a *six* and $P(\bar{S})$ as the probability of *not obtaining* a *six* so that $P(S) = \frac{1}{6}$ and $P(\bar{S}) = \frac{5}{6}$ and $P(\bar{S}) = 1-P(S)$.

In §34, if the selection of an *ace* is a success, $p = \frac{1}{13}$ and $q = \frac{12}{13}$ or $P(A) = \frac{1}{13}$ and $P(\bar{A}) = \frac{12}{13}$.

In §35 $P(H) = \frac{1}{2}$ and $P(\bar{H}) = \frac{1}{2}$ also.

39. Addition rule. Suppose a boy casts a die and that he is to receive a prize if he obtains a *six* or a *one*. The probability of him winning a prize is

$$P(6)+P(1) = \tfrac{1}{6}+\tfrac{1}{6} = \tfrac{1}{3}.$$

This illustrates the *addition rule* for the probabilities of mutually exclusive events which may be stated as follows:

If A and B are MUTUALLY EXCLUSIVE outcomes of a random process, the probability that A or B will occur is the SUM of their probabilities and we write

$$P(A \text{ or } B) = P(A)+P(B),$$

$$\text{or} \quad P(A \cup B) = P(A)+P(B),$$

$$\text{or} \quad P(A+B) = P(A)+P(B).$$

On the other hand, if the two outcomes A and B of a random process are

NOT MUTUALLY EXCLUSIVE the probability that A or B or *both* will occur is the probability of A plus the probability of B minus the probability that both A and B occur and we write

$$P(A \text{ or } B) = P(A)+P(B)-P(AB),$$

$$\text{or} \quad P(A \text{ or } B) = P(A)-P(A \text{ and } B),$$

$$\text{or} \quad P(A \cup B) = P(A)+P(B)-P(A \cap B).$$

Note that the symbols $\cup$ and $\cap$ which are often called *cup* and *cap* respectively may be read simply as OR and AND. A simple example of this last rule is the probability of obtaining a *spade* or an *ace* or *both* (the ace of spades) when a card is drawn at random from a well shuffled pack. This is

$$\begin{aligned} P(S \cup A) &= P(S)+P(A)-P(S \cap A) \\ &= \tfrac{13}{52}+\tfrac{4}{52}-\tfrac{1}{52} \\ &= \tfrac{16}{52}. \end{aligned}$$

This result, of course, can be obtained by arguing from the beginning that 16 of the 52 cards are spades or aces of other suits.

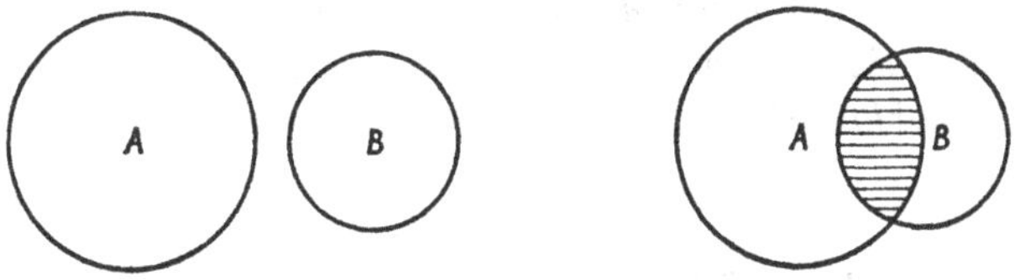

(a) (b)

Fig. 7. (*a*) Outcomes which are mutually exclusive.
(*b*) Outcomes which are NOT mutually exclusive.

The addition rules for probabilities are illustrated diagrammatically by the *Venn diagrams* shown in fig. 7. The probabilities $P(A)$ and $P(B)$ that a bullet fired *at random* will hit either of the targets A and B are proportional to the areas and we may write

$$P(A) = k \times \text{area } A, \quad P(B) = k \times \text{area } B.$$

In case (*a*), the bullet cannot hit both A and B. This is an example of mutually exclusive outcomes and

$$\begin{aligned} P(A \cup B) &= k \times (\text{total area}) \\ &= k \times (\text{area } A + \text{area } B) \\ &= k \times \text{area } A + k \times \text{area } B \\ &= P(A)+P(B). \end{aligned}$$

In case (*b*), the bullet can hit both A and B by falling in the shaded area and the probability of this joint occurrence is

$$P(AB) = P(A \text{ and } B) = P(A \cap B) = k \times \text{shaded area.}$$

But in this case

$$P(A \text{ or } B) = k \times (\text{area } A + \text{area } B - \text{shaded area})$$

because the sum of the areas A and B contains the shaded area *twice*. Hence

$$P(A \text{ or } B) = k \times \text{area } A + k \times \text{area } B - k \times \text{shaded area}$$
$$= P(A) + P(B) - P(AB)$$

which is also written

$$P(A \cup B) = P(A) + P(B) - P(A \cap B).$$

40. Independent events. Suppose a boy casts a die, selects a card from a well-shuffled pack and tosses a coin and that he is to receive a prize only if he obtains a *six* AND an *ace* AND a *head*. The probability of his success is then

$$\tfrac{1}{6} \times \tfrac{1}{13} \times \tfrac{1}{2} = \tfrac{1}{156}.$$

This illustrates the *multiplication rule* for probabilities of *independent events* which states: *If the probabilities of several independent events are* p_1, p_2, ..., p_n, *the probability that* ALL *will occur is the* PRODUCT

$$p_1 p_2 \dots p_n.$$

For the two independent events A and B this is also written

$$P(AB) = P(A)P(B) \quad \text{or} \quad P(A \cap B) = P(A)P(B)$$

and for n independent events $A_1, A_2, \dots, A_n$

$$P(A_1 A_2 \dots A_n) = P(A_1)P(A_2) \dots P(A_n)$$

or

$$P(A_1 \cap A_2 \cap \dots \cap A_n) = P(A_1)P(A_2) \dots P(A_n).$$

41. Conditional probabilities. Consider the following example:

If two cards are drawn from a well-shuffled pack of 52 *playing cards, what is the probability that*:

(i) *they are both aces*;

(ii) *neither of them is an ace*;

(iii) *at least one of them is an ace?*

(i) The probability that the first is an ace is $\tfrac{4}{52}$. If the first is an ace there are only 3 aces left in the remaining 51 cards and the probability that the second is an ace is $\tfrac{3}{51}$. Thus the probability that both are aces is

$$(\tfrac{4}{52}) \times (\tfrac{3}{51}) = \tfrac{1}{221}.$$

Here, we have an example of the rule for *conditional probabilities* which may be stated: *Given two events A and B, the probability of their joint occurrence is the product*

probability of A × conditional probability of B given that A has occurred,

or the product

probability of B × conditional probability of A given that B has occurred.

In symbols, this is written

$$P(AB) = P(A)P(B|A) = P(B)P(A|B),$$

$$\text{or} \quad P(A \cap B) = P(A)P(B|A) = P(B)P(A|B).$$

The symbol $P(B|A)$ may be interpreted as follows: 'If A *has* occurred the probability that B *will* occur is'. Thus, in the example under consideration, the probability that both are aces may be written

$$P(A_1 \cap A_2) = P(A_1)P(A_2|A_1) = (\tfrac{4}{52}) \times (\tfrac{3}{51}) = \tfrac{1}{221}$$

where A_1 means 'the first is an ace' and A_2 means 'the second is an ace'.

Also (ii) the probability that neither is an ace is

$$P(\bar{A}_1 \cap \bar{A}_2) = P(\bar{A}_1)P(\bar{A}_2|\bar{A}_1) = (\tfrac{48}{52})(\tfrac{47}{51}) = \tfrac{188}{221};$$

and (iii) the probability that at least one is an ace is

$$1 - P(\bar{A}_1 \cap \bar{A}_2) = 1 - \tfrac{188}{221} = \tfrac{33}{221}.$$

42. Examples.

1. *If three dice are thrown together, what is the probability of obtaining*

(i) 3 *fives*;
(ii) 1, *and only* 1, *five*;
(iii) *at least* 1 *five?*

(i) By the multiplication rule for independent events, the probability of 3 *fives* is $(\frac{1}{6})^3 = \frac{1}{216}$.

(ii) By §§38 to 40, the first die may be a *five* with the second and third not *or* the second a *five* and the first and third not *or* the third a *five* with the first and second not. Thus the total probability is $3(\frac{1}{6} \times \frac{5}{6} \times \frac{5}{6}) = \frac{25}{72}$.

(iii) The probability of obtaining at least 1 *five*

$$= 1 - \text{the probability of obtaining no } \textit{fives}$$
$$= 1 - (\tfrac{5}{6})^3$$
$$= \tfrac{91}{216}.$$

2. *A box contains ten radio valves all apparently sound, although four of them are actually substandard. Find the chance that, if two of the valves are taken from the box, they are both substandard.* [Northern]

The probability that the first taken from the box is substandard is $\frac{4}{10}$. If the first taken from the box is substandard, the probability that the second is also substandard is $\frac{3}{9}$. Hence the probability that both are substandard is $\frac{4}{10}\times\frac{3}{9} = \frac{2}{15}$. An alternative reasoning for this result is given in §50.

3. *In a cricket match each over consists of six balls and the bowling is opened by two players A and B, the player A bowling the first over. If the probability that A takes a wicket with each ball is $\frac{1}{12}$ and the corresponding probability for B is $\frac{1}{15}$, show that the probability that A takes at least one wicket in his first over is approximately $\frac{2}{5}$ whilst the probability that he takes a wicket before B is approximately $\frac{2}{3}$.* [Northern]

The probability that A does not take a wicket in his first over is $(\frac{11}{12})^6 = 0.5932$. Hence the probability that he takes at least one wicket in his first over is $1-0.5932 = 0.4068$ which is approximately $\frac{2}{5}$.

We have just proved that the probability that A takes his first wicket in his first over is 0.4068. The probability that neither B nor A take wickets in their first overs but that A takes his first wicket in his second over is $(\frac{11}{12})^6(\frac{14}{15})^6\times 0.4068$. The probability that neither B nor A take wickets in their first two overs but that A takes his first wicket in his third over is $(\frac{11}{12})^{12}(\frac{14}{15})^{12}\times 0.4068$, and so on. Thus the infinite geometric progression

$$0.4068+(\tfrac{11}{12})^6(\tfrac{14}{15})^6\times 0.4068+(\tfrac{11}{12})^{12}(\tfrac{14}{15})^{12}\times 0.4068+\ldots$$

gives the total probability that A will take a wicket before B. The sum to infinity of this geometric progression is

$$\frac{0.4068}{1-(\frac{11}{12})^6(\frac{14}{15})^6} = \frac{0.4068}{0.6081}$$

which is approximately $\frac{2}{3}$.

4. *Three per cent of the sparking-plugs manufactured by a firm are defective. Calculate the probability of getting at least one defective plug in a random sample of four.*

Because 3 % are defective we can assume that the probability of any plug selected at random being defective is 0.03 and the probability of it being sound is 0.97. Thus the probability of a random sample of four being all sound is $(0.97)^4 = 0.8855$ and the probability of at least one being defective is $1-0.8855 = 0.1145$ which is approximately $\frac{1}{9}$.

43. Exercises.

1. When three marksmen take part in a shooting contest their chances of hitting the target are $\frac{1}{2}$, $\frac{1}{3}$ and $\frac{1}{4}$. Calculate the chance that one, and only one, bullet will hit the target if all three men fire at it simultaneously. [Northern]

2. The independent probabilities that three components of a television set will need replacing within a year are $\frac{1}{10}$, $\frac{1}{12}$ and $\frac{1}{15}$. Calculate the probability that (i) at least one component, (ii) one and only one component, will need replacing. [Northern]

3. Five per cent of a large consignment of eggs are bad. Find the probability of getting at least one bad egg in a random sample of a dozen. [Northern]

4. A bag contains 7 black balls and 3 white balls. If they are drawn one by one from the bag, find the probability of drawing first a black, then a white and so on alternately until only black balls remain.

5. Three of five dice are each numbered in the normal way, but the fourth is numbered 1, 2, 3, 6, 5, 6 and the fifth 1, 6, 3, 6, 5, 6. If the five dice are thrown together, calculate the probability of turning up (i) five 6's, (ii) at least four 6's, (iii) at least one 6. [Northern]

6. Four players A, B, C, D in this order throw a die in turn. Find for each player the probability of his being first to throw a six. Explain how the sum of the four probabilities forms a useful check. [Northern]

7. In a bag containing 24 marbles there are 4 marbles of each of 6 different colours, white, black, green, red, blue and yellow. Find the probability that 3 marbles drawn together at random from the bag are (i) all white, (ii) all of one colour, (iii) all of different colours. [Northern]

8. A bag contains m oranges and n lemons; sampling is random without replacement.

Obtain the probabilities of the following events:

(i) The first fruit drawn is an orange.
(ii) The first two fruit drawn are oranges.
(iii) The second fruit is a lemon, given that the first was an orange.
(iv) The third fruit is an orange. [Cambridge]

9. Twenty men took equal parts in a rescue operation, and a medal is to be awarded to one of them, chosen at random. Comment on the fairness of the following selection procedures:

(*a*) Twenty pieces of paper, one of which bears a cross, are folded and put in a hat. The men take one at a time, without replacement, until one of them takes the paper with the cross, and this man receives the medal.

(*b*) Each man is allotted a different number from 1 to 20. Two sets of cards numbered from 1 to 10 are shuffled and one card is drawn from each. The man having a number equal to the sum of the two numbers on the cards receives the medal.

(*c*) Each man is allotted a number from 1 to 20 and an independent person is asked to choose a number from 1 to 20 at random. The man with the corresponding number is given the medal.

(*d*) The names of the men are written on twenty similar pieces of paper which are put in a hat and shaken. An independent person takes one of the papers out, and the man named on it receives the medal. [A.E.B.]

10. In a game for two players a turn consists of throwing a die either once or twice, once if the score obtained is less than 6, twice if the score at the first throw is 6. The score for the turn is in the first case the score of the single throw, and in the second case the total score of the two throws.

Obtain the probabilities of a player:

(*a*) scoring more than 9 in a single turn;

(*b*) scoring a total of more than 20 in two succeeding turns;

(*c*) obtaining equal scores in two succeeding turns. [Cambridge]

44. *A posteriori* probabilities. All the previous examples have been concerned with probabilities that estimate the likelihood that an event *will* occur. They are calculated *prior* to observing the results of an experiment and are *a priori*, or *prior*, probabilities. We shall next discuss an example in which a probability is calculated after the outcome of an experiment has been observed. This probability is an example of an *a posteriori*, or posterior, probability.

Suppose that 40 % of the total output of a factory is produced in workshop A and the other 60 % in workshop B. Suppose also that 14 components out of every 1000 produced in A are defective and 6 out of every 1000 produced in B are defective. After the outputs of A and B have been thoroughly mixed a component drawn at random is found to be defective. What is the probability that it is from workshop B?

The probability that the component comes from A

$$P(A) = 0.4.$$

If it comes from A, the probability that it is defective

$$P(D|A) = 0.014.$$

Thus, the probability that it comes from A and is defective

$$P(A \cap D) = P(A)P(D|A) = 0.0056.$$

Similarly, the probability that the component comes from B and is defective

$$P(B \cap D) = P(B)P(D|B) = 0.0036.$$

The ratio of these two probabilities $P(A \cap D):P(B \cap D) = 56:36$, from which we may deduce that if a component drawn from the whole output is found to be defective, the probability that it comes from B

$$P(B|D) = 36/(56+36) = 9/32.$$

It is important to realise that the particular result 9/32 can be generalised as

$$P(B|D) = \frac{P(D|B)P(B)}{P(D|A)P(A)+P(D|B)P(B)}$$

and also

$$P(A|D) = \frac{P(D|A)P(A)}{P(D|A)P(A)+P(D|B)P(B)}.$$

45. Bayes' theorem. The final result of §44 may be extended to three workshops and stated:

If a component drawn at random from the whole output is defective, the probability that it comes from workshop A

$$P(A|D) = \frac{P(D|A)P(A)}{P(D|A)P(A)+P(D|B)P(B)+P(D|C)P(C)},$$

the probability that it comes from workshop B

$$P(B|D) = \frac{P(D|B)P(B)}{P(D|A)P(A)+P(D|B)P(B)+P(D|C)P(C)},$$

and the probability that it comes from workshop C

$$P(C|D) = \frac{P(D|C)P(C)}{P(D|A)P(A)+P(D|B)P(B)+P(D|C)P(C)}.$$

From the consideration of the three workshops and their defectives we may pass on to the relation between n outcomes of a random process $A_1, A_2, \ldots, A_n$ and a chance event B and arrive at *Bayes' theorem* which may be stated:

Let $A_1, A_2, \ldots, A_n$ be a mutually exclusive and exhaustive set of outcomes of a random process, and B be a chance event such that $P(B) \neq 0$, then

$$P(A_r|B) = \frac{P(A_r)P(B|A_r)}{\sum_{r=1}^{n} P(A_r)P(B|A_r)}.$$

This theorem may be proved formally as follows:

Since
$$P(A_r \cap B) = P(A_r)P(B|A_r) = P(B)P(A_r|B)$$

it follows that
$$P(A_r|B) = \frac{P(A_r)P(B|A_r)}{P(B)}.$$

But $P(B)$ is the SUM of all the probabilities of B conditional upon the mutually exclusive and exhaustive set of outcomes $A_1, A_2, \ldots, A_n$. Thus

$$P(B) = P(A_1)P(B|A_1)+P(A_2)P(B|A_2)+\ldots+P(A_n)P(B|A_n)$$

$$= \sum_{r=1}^{n} P(A_r)P(B|A_r).$$

Hence
$$P(A_r|B) = \frac{P(A_r)P(B|A_r)}{\sum_{r=1}^{n} P(A_r)P(B|A_r)}.$$

46. Exercises on Bayes' theorem.

1. In a workshop three machines produce compression springs of a certain size. Machine *A* produces 40 % of the total, machine *B* 25 % and machine *C* the remainder. On the average, 6 % of the springs produced by *A* do not conform to the tolerance requirements. The corresponding percentages for *B* and *C* are 4 and 3 respectively. One spring, selected at random from the whole output, is found to be unsatisfactory. Calculate the probability that it was produced by *A*.

2. Two types of tube, *A* and *B*, are installed in a certain make of television set three-quarters of the tubes being of type *A*. The probability that type *A* will fail within a year is $\frac{1}{8}$ and the probability that type *B* will fail within a year is $\frac{1}{4}$. If a set fails within a year, what is the probability that it contained a *B* tube?

3. Two bags contain different numbers of black and white balls. Bag *A* contains 3 white and 7 black, bag *B* contains 6 white and 4 black. If the result of tossing a coin is a *head* a ball is drawn from *A* but if the result is a *tail* a ball is drawn from *B*. The ball drawn is white. Calculate the posterior probability that the result of the toss was a *head*.

4. Box *A* contains 3 white and 3 black balls and box *B* contains 4 white and 3 black balls. One ball is transferred from *A* to *B*. One ball is then drawn from *B* and is found to be white. What is the probability that the transferred ball was white?

5. A single die is cast. If a 1 results, a ball is drawn from urn *A* which contains 1 white and 4 black balls; if 2 or 3 results, a ball is drawn from urn *B* which contains 2 white and 3 black balls; if 4, 5 or 6 result, a ball is drawn from urn *C* which contains 3 white and 2 black balls. The experiment is carried out and results in a white ball being drawn. Calculate the probability that the result of the throw was a 1.

6. For salesmen marketing a certain article two methods of approach, *A* and *B*, may be made to customers. If the new method *A* is used the probability of success is $\frac{3}{4}$ but if the old method *B* is used the probability of success is only $\frac{1}{2}$. Twenty-five per cent of the salesmen have adopted method *A* but the remainder still use method *B*. If one of the sales successes is selected at random from the whole list of successes, what is the probability that it was the result of approach *A*?

7. When three candidates, *A*, *B* and *C*, put up for Council, it was estimated that their respective probabilities of being elected were $\frac{1}{8}$, $\frac{1}{2}$ and $\frac{3}{8}$. It was also estimated that, if either *A* or *C* were elected, the probability of a new Public Library being built was $\frac{1}{2}$ but if *B* were elected the corresponding probability was $\frac{1}{4}$. There is now a new Public Library. What is the probability that *A* was elected?

4

PROBABILITY DISTRIBUTIONS

47. The symbol $\binom{n}{r}$ or ^nC_r. The symbol $\binom{n}{r}$ or ^nC_r denotes the number of ways of *choosing* (or *selecting*) r things from n unlike things. It is alternatively stated to be the number of *combinations* (or *selections*) of n unlike things taken r at a time and it is proved in textbooks of Algebra that

$$\binom{n}{r} = \frac{n(n-1)(n-2)\ldots(n-r+1)}{1.2.3\ldots r}.$$

Thus, the number of ways of choosing a sample of 4 radio valves from a box of 10 is

$$\binom{10}{4} = \frac{10.9.8.7}{1.2.3.4}$$
$$= 210$$

and if the 4 valves are *chosen at random* each of these 210 ways are equally probable.

Again, the number of ways of choosing 3 cards from an ordinary pack of 52 playing cards is

$$\binom{52}{3} = \frac{52.51.50}{1.2.3}$$
$$= 22100.$$

48. The factorial notation $n!$ or $\underline{|n}$. The symbol $n!$ or $\underline{|n}$ is used to denote the product of the first n positive integers. Thus

$$n! = 1.2.3\ldots n.$$

For example $\quad 7! = 1.2.3.4.5.6.7 = 5040,$

and $\quad 4! = 1.2.3.4 = 24.$

An alternative method of writing

$$\binom{10}{4} = \frac{10.9.8.7}{1.2.3.4}$$

is, therefore,

$$\binom{10}{4} = \frac{10.9.8.7}{1.2.3.4} \times \frac{6.5.4.3.2.1}{1.2.3.4.5.6}$$
$$= \frac{10!}{4!6!}$$

and in general it is customary to write

$$\binom{n}{r} = \frac{n!}{r!(n-r)!}.$$

49. Selecting and rejecting. The number of ways of selecting 11 players from the 14 members of a cricketing party is

$$\binom{14}{11} = \frac{14.13.12.11.10.9.8.7.6.5.4}{1.2.3.4.5.6.7.8.9.10.11}$$
$$= 364.$$

Clearly, this is more conveniently calculated by using the fact that the number of ways of selecting the 11 players from the 14 members is the same as the number of ways of selecting the 3 non-players which is

$$\binom{14}{3} = \frac{14.13.12}{1.2.3}$$
$$= 364.$$

Thus $$\binom{14}{11} = \binom{14}{3}$$

and in general $$\binom{n}{r} = \binom{n}{n-r}$$

The last equation is stated in words as follows: *The number of ways of selecting r things from n unlike things is the same as the number of ways of rejecting $(n-r)$.*

50. The use of $\binom{n}{r}$ in examples on probability. In §41, it was shown that if 2 cards are selected at random from a well-shuffled pack of 52 playing cards, the probability that they are both aces is $\frac{1}{221}$ whilst the probability that neither are aces is $\frac{188}{221}$. An alternative argument which leads to these same values is as follows:

(i) The probability that both are aces

$$= \frac{\text{the number of ways of selecting 2 aces from the 4 in the pack}}{\text{the number of ways of selecting any 2 of the 52 cards in the pack}}$$

$$= \binom{4}{2} \Big/ \binom{52}{2}$$

$$= \frac{4.3}{1.2} \Big/ \frac{52.51}{1.2}$$

$$= \tfrac{1}{221}.$$

(ii) The probability that neither are aces

$$= \frac{\text{the number of ways of selecting 2 cards from the 48 which are not aces}}{\text{the number of ways of selecting any 2 of the 52 cards}}$$

$$= \binom{48}{2} \Big/ \binom{52}{2}$$

$$= \frac{48.47}{1.2} \Big/ \frac{52.51}{1.2}$$

$$= \tfrac{188}{221}.$$

The result of §42, Ex. 2, might be established by a similar argument as follows:

The probability that both valves are substandard is

$$= \frac{\text{the number of ways of choosing any 2 of the 4 substandard valves}}{\text{the number of ways of choosing any 2 of the 10 valves}}$$

$$= \binom{4}{2} \Big/ \binom{10}{2}$$

$$= \frac{4.3}{1.2} \Big/ \frac{10.9}{1.2}$$

$$= \tfrac{2}{15}.$$

51. Experiments with playing cards.

Experiment 1. *Suppose we draw at random 5 cards from a well-shuffled pack and note the number of spades.*

(i) The probability that *none* are spades is

$$\binom{39}{5} \Big/ \binom{52}{5} = 0.2215.$$

(ii) The probability that 1 is a spade and 4 are not is calculated as follows:

The number of ways of selecting 4 which are not spades is $\binom{39}{4}$.

The number of ways of selecting 1 which is a spade is $\binom{13}{1}$.

Each selection of 4 which are not spades can be combined with each selection of 1 which is a spade to form a different choice of 5 cards and hence

$$\binom{39}{4} \times \binom{13}{1}$$

is the complete number of ways of selecting 5 cards, 4 of which are not spades and 1 of which is a spade.

Thus, the probability that 1 is a spade and 4 are not is

$$\binom{39}{4} \times \binom{13}{1} \Big/ \binom{52}{5} = 0.4114.$$

Table 4A summarizes all possible cases. It is an example of a *probability distribution.* To test the theory experimentally the student should take a pack of cards and

(i) shuffle it thoroughly,
(ii) draw out 5 cards at random,
(iii) note how many of the 5 are spades,
(iv) replace the 5 cards.

If this cycle of four operations is repeated 50 times, the observed frequencies thus obtained should approximate quite closely to the probable frequencies of table 4A obtained by multiplying the probabilities by 50.

TABLE 4A

Number of spades in the random sample of 5	Probability	Probable frequency for 50 random samples
0	$\binom{39}{5} \Big/ \binom{52}{5} = 0.2215$	11
1	$\binom{39}{4} \binom{13}{1} \Big/ \binom{52}{5} = 0.4114$	20
2	$\binom{39}{3} \binom{13}{2} \Big/ \binom{52}{5} = 0.2743$	14
3	$\binom{39}{2} \binom{13}{3} \Big/ \binom{52}{5} = 0.0816$	4
4	$\binom{39}{1} \binom{13}{4} \Big/ \binom{52}{5} = 0.0107$	1
5	$\binom{13}{5} \Big/ \binom{52}{5} = 0.0005$	0

Experiment 2. *Suppose we draw at random* 5 *cards from a well-shuffled pack and note the number of aces.* Similar arguments to those of Experiment 1 enable us to establish table 4B which should again be tested experimentally. It provides a second example of a probability distribution.

52. The number of ways of selecting like things. It will be realised that the probabilities shown in table 4A apply equally well to a random sample of 5 balls drawn from a bag containing 13 black balls and 39 balls of other colours, or to a random sample of 5 cycle lamp batteries taken from a box containing 52 batteries 13 of which are defective. In this connection it is important to note that the number of ways of selecting 5 black balls from 13 which are all alike is $\binom{13}{5} = 1287$ even though it is impossible to tell the difference between these 1287 selections. Thus, although only one selection is apparent, there are 1287 ways of making it. Probabilities are calculated from the number of ways in which selections can be made. The examples which follow should help to make clear the distinction between 'the number of selections' and 'the number of ways of selecting'.

TABLE 4B

Number of aces in the random sample of 5	Probability	Probable frequency for 50 random samples
0	$\binom{48}{5}\Big/\binom{52}{5} = 0.6589$	33
1	$\binom{48}{4}\binom{4}{1}\Big/\binom{52}{5} = 0.2995$	15
2	$\binom{48}{3}\binom{4}{2}\Big/\binom{52}{5} = 0.0399$	2
3	$\binom{48}{2}\binom{4}{3}\Big/\binom{52}{5} = 0.0017$	0
4	$\binom{48}{1}\binom{4}{4}\Big/\binom{52}{5} = 0.0000\ (2)$	0

53. Examples.

1. (i) If 3 letters are chosen from the word SWEETER how many different selections are possible?

(ii) What is the probability that a random selection of 3 letters from the same word contains at least 2 E's?

(i) If 3 E's are chosen, only 1 selection is possible.

If 2 E's and 1 other letter are chosen, 4 selections are possible.

If 1 E and 2 other letters are chosen, $\binom{4}{2} = 6$ selections are possible.

If no E's and 3 other letters are chosen, $\binom{4}{3} = 4$ selections are possible.

Hence the total number of different selections is 15.

(ii) The probability that the selection contains 3 E's

$$= \frac{\text{the no. of ways of selecting 3 E's}}{\text{the no. of ways of selecting 3 letters from 7}}$$

$$= 1\Big/\binom{7}{3}$$

$$= \tfrac{1}{35}.$$

The probability that the selection contains 2 E's

$$= \frac{\text{the no. of ways of combining 2 E's from 3 with 1 other letter from 4}}{\text{the no. of ways of selecting 3 letters from 7}}$$

$$= \frac{\binom{3}{2}\times\binom{4}{1}}{\binom{7}{3}}$$

$$= \tfrac{12}{35}.$$

Hence the probability that a random selection of 3 letters contains at least 2 E's is $\frac{13}{35}$.

2. Three balls are drawn at random from a bag containing 3 red, 4 white and 5 black balls. Calculate the probabilities that the 3 balls are (i) all black, (ii) 1 red, 1 white, 1 black. [Northern]

(i) The probability that the 3 balls are all black

$$= \frac{\text{the no. of ways of selecting 3 balls from 5 black}}{\text{the no. of ways of selecting 3 balls from 12}}$$

$$= \binom{5}{5}\Big/\binom{12}{3}$$

$$= \tfrac{1}{22}.$$

(ii) The probability that 1 is red, 1 is black and 1 is white

$$= \binom{3}{1}\binom{4}{1}\binom{5}{1}\Big/\binom{12}{3}$$

$$= \tfrac{3}{11}.$$

3. A department in a works has 10 machines which may need adjustment from time to time during the day. Three of these machines are old, each having a probability of $\frac{1}{11}$ of needing adjustment during the day, and 7 are new, having corresponding probabilities of $\frac{1}{21}$.

Assuming that no machine needs adjustment twice on the same day, determine the probabilities that on a particular day

(i) just 2 old and no new machines need adjustment,

(ii) if just 2 machines need adjustment, they are of the same type.

[Northern]

(i) The probability that 2 of the old machines need adjustment but that the third does not, is $(\frac{1}{11})^2(\frac{10}{11})$.

The 2 old machines which need adjustment can be selected, however, in $\binom{3}{2}$ ways.

Thus the *total* probability that just two of the 3 old machines need adjustment is

$$\binom{3}{2}\left(\frac{1}{11}\right)^2\left(\frac{10}{11}\right).$$

Further, the probability that no new machines need adjustment is $(\frac{20}{21})^7$.

Hence, the *compound* probability that just 2 old and no new machines need adjustment is

$$\binom{3}{2}\left(\frac{1}{11}\right)^2\left(\frac{10}{11}\right)\left(\frac{20}{21}\right)^7 = 0.0160.$$

The student might find the following argument easier.

Let the old machines be X, Y, Z and the new machines A, B, C, D, E, F, G.

The probability that X, Y need adjustment and Z, A, B, C, D, E, F, G do not is $(\frac{1}{11})(\frac{1}{11})(\frac{10}{11})(\frac{20}{21})^7$.

Similarly X, Z may need adjustment but not Y, A, B, C, D, E, F, G, and also Y, Z may need adjustment but not X, A, B, C, D, E, F, G.

Thus the total probability, as before, is $3(\frac{1}{11})^2(\frac{10}{11})(\frac{20}{21})^7$.

(ii) The probability that just 2 new machines and no old machines need adjustment is, similarly,

$$\binom{7}{2}\left(\frac{1}{21}\right)^2\left(\frac{20}{21}\right)^5\left(\frac{10}{11}\right)^3 = 0.0280.$$

'The probability that if just 2 machines need adjustment, they are of the same type' is the same as 'the probability that *either* just 2 old and no new *or* just 2 new and no old need adjustment'. Thus the required probability is

$$0.0160+0.0280 = 0.0440.$$

54. Exercises.

1. Each of two bags contains 8 coins. How many different combinations of 6 coins can be made by drawing 3 coins out of each bag, (i) when no two of the 16 coins are alike, (ii) when two of the coins in one bag are alike?

In case (i) what is the probability that a specified coin will appear in any one combination? [London]

2. In question 1, case (ii) above, what is the probability that at least one of the two like coins will appear in any one combination?

3. A box contains red and green sweets mixed together in the ratio 3 to 2. A handful of 20 sweets is taken at random from the box. Find

(i) the probability that there will be precisely 15 red sweets in the handful,

(ii) the probability that the handful of sweets is such that it can be shared equally between 5 children so that each child receives the same number of red sweets, this number being at least 2. [Northern]

4. In an agricultural experiment barley is to be grown on 8 plots of land arranged in a line, with original fertility measurements as shown in fig. 8:

Plot no.	1	2	3	4	5	6	7	8
Fertility	10	10	11	11	11	11	10	10

Fig. 8

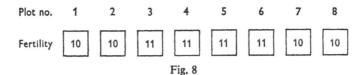

A manurial treatment is applied to 4 of the plots chosen at random. Find the probability that it is applied to precisely 3 plots having a fertility measurement of 11 and one plot having a fertility measurement of 10. Calculate the mean original fertility of manured plots minus the mean original fertility of unmanured plots in this case.

By considering other cases too find the probability distribution of the mean original fertility of manured plots minus the mean original fertility of unmanured plots. [Northern]

5. State the addition and multiplication laws of probability.

A number is chosen at random from each of the two sets

$$0,\ 1,\ 2,\ 3,\ 4,\ 5,\ 6,\ 7,\ 8,\ 9;$$
$$0,\ 1,\ 2,\ 3,\ 4,\ 5,\ 6,\ 7,\ 8,\ 9.$$

Calculate the probability that

(i) both numbers are odd;

(ii) their sum is not greater than 5. [Northern]

6. A stack of 20 cards contains 4 cards of each of 5 different colours, namely white, black, green, red and blue. Apart from colour the cards are indistinguishable. A set of 3 cards is drawn at random from the stack. Find the chances that the 3 cards are (i) all white, (ii) all of one colour, (iii) all of different colours. [Cambridge]

5

THE BINOMIAL DISTRIBUTION

55. Throwing six dice. Suppose 6 dice are thrown together.

(i) The probability, $P(0)$, that *no* sixes fall uppermost is $(\frac{5}{6})^6$.

(ii) The probability, $P(1)$, that *any one* of the dice has its six uppermost but the *other five* have not, is calculated as follows:

The probability that 5 of the dice are not sixes but that the other is a six is $(\frac{5}{6})^5(\frac{1}{6})$. There are, however, $\binom{6}{1}$ ways of selecting the die which is a six and each of these $\binom{6}{1}$ ways are equally probable. Thus the total probability that any 1 is a six and the other 5 are not is

$$\binom{6}{1}\left(\frac{5}{6}\right)^5\left(\frac{1}{6}\right).$$

(iii) Similarly, the probability, $P(2)$, that *any two* of the dice have their sixes uppermost and the *other four* have not is

$$\binom{6}{2}\left(\frac{5}{6}\right)^4\left(\frac{1}{6}\right)^2.$$

Table 5A summarises all possible cases. It shows also the probable frequencies if the six dice are thrown together 216 times.

The following distribution was obtained in an actual experiment:

No. of sixes	0	1	2	3	4	5	6	Total
Frequency	77	84	34	19	2	0	0	216

The student will notice that the *observed frequencies* of the experiment approximate quite closely to the *expected frequencies* forecast in table 5A. It is essential that the beginner should obtain his own set of observed frequencies by carrying out the experiment himself.

56. The binomial expansion of $(q+p)^n$. The probabilities in table 5A are the terms of the binomial expansion

$$(q+p)^6 = q^6+\binom{6}{1}q^5p+\binom{6}{2}q^4p^2+\binom{6}{3}q^3p^3+\binom{6}{4}q^2p^4+\binom{6}{5}qp^5+p^6,$$

where $q = \frac{5}{6}$ and $p = \frac{1}{6}$. It will now be realised that

THE TERMS OF THE GENERAL BINOMIAL EXPANSION

$$(q+p)^n = q^n + \binom{n}{1} q^{n-1}p + \binom{n}{2} q^{n-2}p^2 + \dots + p^n,$$

WHERE $q = \frac{5}{6}$ AND $p = \frac{1}{6}$ GIVE THE PROBABILITIES OF OBTAINING 0, 1, 2, ..., n SIXES WHEN n DICE ARE THROWN TOGETHER.

TABLE 5A

No. of sixes x	Probability $P(x)$	Probable frequency for 216 throws $216.P(x)$
0	$P(0) = \left(\frac{5}{6}\right)^6$	72.34
1	$P(1) = \binom{6}{1}\left(\frac{5}{6}\right)^5\left(\frac{1}{6}\right)$	86.81
2	$P(2) = \binom{6}{2}\left(\frac{5}{6}\right)^4\left(\frac{1}{6}\right)^2$	43.40
3	$P(3) = \binom{6}{3}\left(\frac{5}{6}\right)^3\left(\frac{1}{6}\right)^3$	11.57
4	$P(4) = \binom{6}{4}\left(\frac{5}{6}\right)^2\left(\frac{1}{6}\right)^4$	1.74
5	$P(5) = \binom{6}{5}\left(\frac{5}{6}\right)\left(\frac{1}{6}\right)^5$	0.14
6	$P(6) = \left(\frac{1}{6}\right)^6$	0.00
Total	1	216.00

57. Exercises.

1. Use the binomial expansion of $(q+p)^3$ to calculate the probabilities of obtaining 0, 1, 2, 3 *ones* (or *aces*) when 3 dice are thrown together. Estimate the frequencies that can be expected if the 3 dice are thrown together 216 times and display your results in a table similar to 5A. Test by experiment.

2. Calculate the probabilities of obtaining 0, 1, 2, 3, 4, 5 *threes* when 5 dice are thrown together. Estimate the frequencies that can be expected if the 5 dice are thrown together 100 times and test by experiment.

58. Tossing five coins. The argument of §55 can easily be modified to show that the terms of the binomial expansion

$$(q+p)^5 = p^5+\binom{5}{1}q^4p+\binom{5}{2}q^3p^2+\binom{5}{3}q^2p^3+\binom{5}{4}qp^4+p^5,$$

where $q = \frac{1}{2}$ and $p = \frac{1}{2}$, give the probabilities of obtaining 0, 1, 2, 3, 4, 5 *heads* when 5 coins are tossed together. These probabilities together with the expected frequencies for a total of 832 tosses and the observed frequencies of an actual experiment are displayed in table 5B. The student will find it interesting to carry out the experiment himself but not necessarily for such a large number of tosses.

TABLE 5B

No. of heads x	Probability $P(x)$	Expected frequency for 832 tosses $832.P(x)$	Observed frequency of an actual experiment
0	$P(0) = \left(\frac{1}{2}\right)^5$	26	25
1	$P(1) = \binom{5}{1}\left(\frac{1}{2}\right)^4\left(\frac{1}{2}\right)$	130	123
2	$P(2) = \binom{5}{2}\left(\frac{1}{2}\right)^3\left(\frac{1}{2}\right)^2$	260	246
3	$P(3) = \binom{5}{3}\left(\frac{1}{2}\right)^2\left(\frac{1}{2}\right)^4$	260	264
4	$P(4) = \binom{5}{4}\left(\frac{1}{2}\right)\left(\frac{1}{2}\right)^4$	130	146
5	$P(5) = \left(\frac{1}{2}\right)$	26	28
Total	1	832	832

59. Exercise.

Calculate the probabilities of obtaining 0, 1, 2, 3, 4 *heads* when 4 coins are tossed together. Estimate the frequencies that can be expected if the 4 coins are tossed together 40 times and test by experiment.

60. The sampling-bottle. The *sampling-bottle* shown in fig. 9 contains a large number of small steel balls 5% of which are painted black. If one ball is taken at random from the bottle the probability, p, that it is black is $\frac{1}{20}$ or 0.05 and the probability, q, that it is not black is $\frac{19}{20}$ or 0.95.

Suppose a random sample of 10 balls is taken from the bottle. The probabilities that the sample contains 0, 1, 2, etc., black balls is given by the terms of the binomial expansion

$$(q+p)^{10} = q^{10}+\binom{10}{1}q^9p+\binom{10}{2}q^8p^2+\ldots,$$

where $q = 0.95$ and $p = 0.05$. These probabilities are shown in table 5C. The separate probabilities of 3, 4, 5, etc., black balls appearing in any random sample are so small that they have been combined as *the probability of* 3 *or more*, $P(x \geqslant 3)$, where

$$P(x \geqslant 3) = 1-\{P(0)+P(1)+P(2)\}.$$

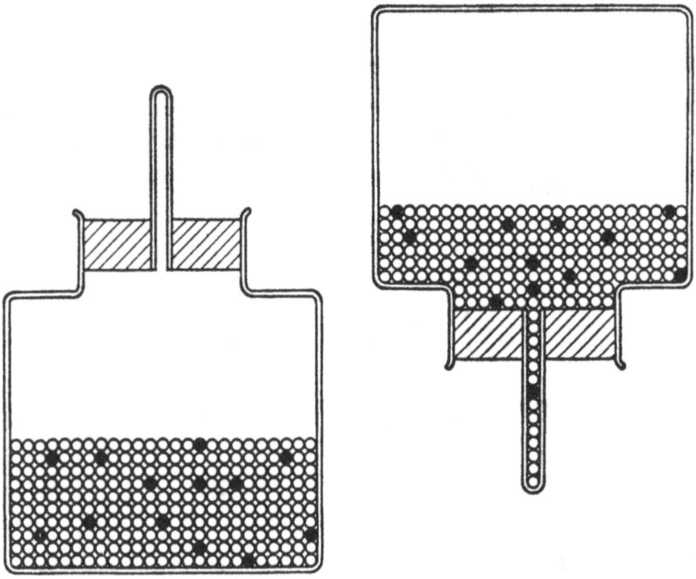

Fig. 9. A sampling-bottle. *Upright position*: the bottle contains a large number of small steel balls 5 % of which are painted black; a hollow glass tube sealed at the outer end and open at the inner end passes through the cork. *Inverted position*: after being well shaken the bottle is turned over and some of the steel balls fall into the hollow glass tube, which is just long enough to allow 10 balls to be visible below the cork.

It will be appreciated that the hollow glass tube passing through the cork is a very convenient arrangement for drawing random samples from the bottle. The process of shaking the bottle, turning it over, counting the number of black balls visible below the cork, is so quick that it can be repeated a hundred times in a few minutes.

61. Exercises.

1. Draw up a table similar to 5c for a sampling-bottle if 10% of the balls contained in it are black and the glass tube is just long enough to draw random samples of five.

2. Draw up a table similar to 5c for a sampling-bottle if $2\frac{1}{2}$% of the balls contained in it are black and the glass tube is long enough to draw samples of twenty.

TABLE 5c

No. of black balls in the random sample of 10 x	Probability $P(x)$	Expected frequency for 100 samples $100.P(x)$
0	$P(0) = (0.95)^{10} = 0.598$	59.8
1	$P(1) = \binom{10}{1}(0.95)^9(0.05) = 0.315$	31.5
2	$P(2) = \binom{10}{2}(0.95)^8(0.05)^2 = 0.075$	7.5
3 or more	$P(x \geqslant 3) = 1-\{P(0)+P(1)+P(2)\} = 0.012$	1.2
Total	1	100.0

62. Sampling in industry. The experiments carried out with the sampling-bottle lead directly to the method of sampling as it is applied in the testing of mass-produced articles which are either accepted or rejected according as to whether or not they satisfy specification requirements. If, for example, 5% of the screws manufactured by a certain process are rejected because they fail to satisfy tolerance requirements, table 5c gives the probabilities of a random sample of 10 screws containing 0, 1, 2, 3 or more rejects. Also, if 100 samples are inspected the frequencies for 0, 1, 2, 3 or more rejects per sample should approximate fairly closely to the expected frequencies of table 5c, any large divergence signifying that the proportion of rejects in the bulk differs significantly from 5%. The examples and exercises which follow will serve to make clear the important practical applications of the binomial distribution.

63. Examples.

1. (i) *In a large lot of electric-light bulbs 5% of the bulbs are defective. Calculate the probability that a random sample of 20 will contain at most 2 defective bulbs.*

(ii) *One third of the lots presented for inspection have 5 % defective, the rest 10 % defective. If a lot is rejected when a random sample of 20 taken from it contains more than 2 defective bulbs, find the proportion of lots which are rejected.* [Northern]

(i) Using the terms of the binomial expansion

$$(q+p)^{20} = q^{20}+\binom{20}{1}q^{19}p+\binom{20}{2}q^{18}p^2+\ldots$$

with $p = 0.05$ and $q = 0.95$ it is convenient to tabulate the calculation in a form similar to table 5C, as shown in table 5D.

Thus, the probability that a random sample of 20 will contain at most 2 defective bulbs is 0.9236. This means that for lots which have 5 % defective about 92 % of the samples will have at most 2 defective bulbs and 8 % of the samples will have more than 2 defective bulbs.

TABLE 5D

No. of defective bulbs in random sample of 20 x	Probability $P(x)$
0	$P(0) = (0.95)^{20} = 0.3581$
1	$P(1) = \binom{20}{1}(0.95)^{19}(0.05) = 0.3770$
2	$P(2) = \binom{20}{2}(0.95)^{18}(0.05)^2 = 0.1885$
At most 2	$P(x \leqslant 1) = P(0)+P(1)+P(2) = 0.9236$

TABLE 5E

No. of defective bulbs in random sample of 20 x	Probability $P(x)$
0	$P(0) = (0.9)^{20} = 0.1213$
1	$P(1) = \binom{20}{1}(0.9)^{19}(0.1) = 0.2696$
2	$P(2) = \binom{20}{2}(0.9)^{18}(0.1)^2 = 0.2846$
At most 2	$P(x \leqslant 2) = P(0)+P(1)+P(2) = 0.6755$

(ii) Table 5E gives values for lots which have 10 % defective $p = 0.1$ and $q = 0.9$.

Thus, for lots which have 10 % defective about 68 % of the samples will have at most 2 defective bulbs and 32 % of the samples will have more than 3 defective bulbs.

Let us suppose that 300 lots are presented for inspection, 100 of them having 5 % defective and 200 having 10 % defective. Then 8 of the 100 and 2×32 of the 200 will be rejected.

Thus 72 of the 300 lots will be rejected.

This means that 24 % of all the lots presented for inspection will be rejected.

2. *Assuming that the probability that any one of* 8 *telephone lines is engaged at an instant is* $\frac{1}{4}$, *calculate the probability that*

(i) *at least one of the lines is engaged,*
(ii) *all* 8 *lines are engaged.*

What is the most probable number of lines engaged at any instant and what is the probability that this number of lines is engaged?

The simplest method of calculation is by a table similar to 5C (table 5F).

TABLE 5F

No. of lines engaged x	Probability $P(x)$
0	$P(0) = (\frac{3}{4})^8 = 0.1002$
1	$P(1) = \binom{8}{1}\left(\frac{3}{4}\right)^7\left(\frac{1}{4}\right) = 0.2669$
2	$P(2) = \binom{8}{2}\left(\frac{3}{4}\right)^6\left(\frac{1}{4}\right)^2 = 0.3114$
3	$P(3) = \binom{8}{3}\left(\frac{3}{4}\right)^5\left(\frac{1}{4}\right)^3 = 0.2076$
⋮	⋮
8	$P(8) = (\frac{1}{4})^8 = 0.00001524$

The probability that at least one line is engaged is

$$P(1)+P(2)+P(3)+\ldots+P(8) = 1-P(0)$$
$$= 0.8998,$$

which is approximately $\frac{9}{10}$.

The probability that all the lines are engaged is 0.00001524. The most probable number of lines engaged at any instant is 2 and the probability of this number of lines being engaged is 0.3114 (rather less than $\frac{1}{3}$).

A more direct way of finding the most probable number is as follows:

The probabilities $P(0)$, $P(1)$, $P(2)$, etc., continue to increase provided

$$\frac{P(x+1)}{P(x)} \geqslant 1.$$

Now
$$\frac{P(x+1)}{P(x)} = \frac{\binom{8}{x+1}\left(\frac{3}{4}\right)^{7-x}\left(\frac{1}{4}\right)^{x+1}}{\binom{8}{x}\left(\frac{3}{4}\right)^{8-x}\left(\frac{1}{4}\right)^{x}}$$

$$= \left\{\frac{8!}{(x+1)!(7-x)!}\Big/\frac{8}{x!(8-x)!}\right\}\frac{(\frac{1}{4})}{(\frac{3}{4})}$$

$$= \frac{(8-x)}{3(x+1)}.$$

Thus $\dfrac{P(x+1)}{P(x)} \geqslant 1$ provided $(8-x) \geqslant 3(x+1)$.

This gives $x \leqslant \frac{5}{4}$ and hence $P(2)$ is the greatest probability.

64. Exercises.

1. In the manufacture of screws by a certain process it was found that 8 % of the screws were rejected because they failed to satisfy tolerance requirements. What was the probability that a sample of 10 screws contained (i) exactly 2, (ii) not more than 2 rejects?

2. Fifty samples of 20 are drawn from a large bulk in which the proportion of defective items is 12 %. Calculate the number of samples that can be expected to include 0, 1, 2, etc., defectives.

3. If on the average rain falls on 12 days in every 30 find the probability (i) that the first 3 days of a given week will be fine and the remainder wet; (ii) that rain will fall on just 3 days of a given week. [London]

4. A marksman's chance of hitting a target with each of his shots is $\frac{2}{5}$. If he fires 6 shots calculate his chance of (i) exactly 3 hits, (ii) at least 3 hits.

5. In a cricket match the probability that a certain bowler takes a wicket with any ball is $\frac{1}{15}$. Calculate the probability of him taking (i) 3 wickets in 3 balls, (ii) at least 1 wicket in 6 balls.

6. The probability that a tennis player will serve an 'ace' is $\frac{1}{5}$. What is the probability that she will serve exactly 4 aces out of 7 services?

65. Verification that the mean of the binomial distribution is *np*.

(i) From table 5A, the mean number of sixes per throw,

$$\bar{x} = \tfrac{1}{216}\{72.34\times 0+86.81\times 1+43.40\times 2+11.57\times 3+1.74\times 4+0.14\times 5\}$$

$$= 1.$$

(ii) From table 5B, the mean number of heads per toss,

$$\bar{x} = \tfrac{1}{832}\{26\times 0+130\times 1+260\times 2+260\times 3+130\times 4+26\times 5\}$$

$$= 2\tfrac{1}{2}.$$

(iii) From table 5C it is only possible to obtain an approximate value for the mean number of black balls per sample because of the combined probability for '3 or more'. This approximate value

$$\bar{x} = \tfrac{1}{100}\{59.8\times 0+31.5\times 1+7.5\times 2+1.2\times 3\}$$

$$= 0.501.$$

The above three values (i), (ii) and (iii) of $\bar{x}$ are examples of the following valuable property of the binomial distribution which will be proved formally in §66.

If a variate x takes the values 0, 1, 2, ..., n with probabilities equal respectively to the terms, arranged in the usual way, of the binomial expansion of $(q+p)^n$, where $q+p = 1$, then the mean value of x is np.

Thus (i) in table 5A, $n = 6$, $p = \frac{1}{6}$ and the mean $np = 1$,
(ii) in table 5B, $n = 5$, $p = \frac{1}{2}$ and the mean $np = 2\frac{1}{2}$,
and (iii) in table 5C, $n = 10$, $p = 0.05$ and the mean $np = 0.5$.

66. The mean and variance of the binomial distribution. By setting down the binomial distribution as shown in table 5G its mean and variance can easily be deduced. In theoretical work of this kind the probabilities of column (ii) are often called *relative frequencies.*

TABLE 5G

(i)	(ii)	(iii)	(iv)
Variate x	Probability or relative frequency f	fx (i) × (ii)	fx^2 (i) × (iii)
0	q^n	0	0
1	$nq^{n-1}p$	$nq^{n-1}p$	$nq^{n-1}p$
2	$\frac{n(n-1)}{1.2}q^{n-2}p^2$	$\frac{n(n-1)}{1}q^{n-2}p^2$	$\frac{2n(n-1)}{1}q^{n-2}p^2$
3	$\frac{n(n-1)(n-2)}{1.2.3}q^{n-3}p^3$	$\frac{n(n-1)(n-2)}{1.2}q^{n-3}p^3$	$\frac{3n(n-1)(n-2)}{1.2}q^{n-3}p^3$
⋮	⋮	⋮	⋮
n	p^n	np^n	n^2p^2
Total	Σf	Σfx	Σfx^2

Now $\Sigma f = 1$,

$$\Sigma fx = np\left\{q^{n-1}+\frac{(n-1)}{1}q^{n-2}p+\frac{(n-1)(n-2)}{1.2}q^{n-3}p^2+\ldots+p^{n-1}\right\}$$

$$= np(q+p)^{n-1}$$

$$= np,$$

and
$$\Sigma fx^2 = np\left\{q^{n-1}+\frac{2(n-1)}{1}q^{n-2}p+\frac{3(n-1)(n-2)}{1.2}q^{n-3}p^2+\ldots+np^{n-1}\right\}$$

$$= np\left\{q^{n-1}+\frac{(n-1)}{1}q^{n-2}p+\frac{(n-1)(n-2)}{1.2}q^{n-3}p^2+\ldots+p^{n-1}\right.$$
$$\left.+\frac{(n-1)}{1}q^{n-2}p+\frac{2(n-1)(n-2)}{1.2}q^{n-3}p^2+\ldots+(n-1)p^{n-1}\right\}$$

$$= np\{(q+p)^{n-1}+(n-1)p(q+p)^{n-2}\}$$

$$= np\{1+(n-1)p\}$$

$$= np(1-p)+(np)^2.$$

Thus the mean,
$$\frac{\Sigma fx}{\Sigma f} = np$$

and the variance,
$$\frac{\Sigma fx^2}{\Sigma f}-\left(\frac{\Sigma fx}{\Sigma f}\right)^2 = npq.$$

67. How to estimate the proportion of black balls in a sampling-bottle. Let us now suppose that the proportion of black balls in a sampling-bottle is unknown and that 100 samples of 8 drawn from the bottle give the following distribution:

No. of black balls in the sample of 8	x	0	1	2	3	4	5 or more	Total
No. of samples		32	41	20	6	1	0	100

The mean number of black balls per sample

$$\bar{x} = \tfrac{1}{100}\{32\times0+41\times1+20\times2+6\times3+1\times4\}$$

$$= 1.03,$$

and in this case $n = 8$.

Hence $\bar{x} = np$ gives $1.03 = 8p$ and thus

$$p = 0.13 \text{ approximately.}$$

This means that 13 % of the balls in the bottle are black. Here then is a valuable method of estimating the proportion p of defective items in a bulk by examining random samples of n items.

68. Estimation of the proportion of defective items in a bulk. *The items produced by a certain machine were checked by examining samples of* 12. *The following table shows the distribution of* 50 *samples according to the number of defective items they contained*:

No. of defectives in a sample of 12	0	1	2	3	4	5 or more	Total
No. of samples	18	19	9	3	1	0	50

Calculate the mean number of defectives per sample and, assuming that the binomial law applies, estimate the proportion of defective items in the whole bulk produced by the machine.

By an easy calculation the mean number of defectives per sample is 1, and since $n = 12$, it follows that $p = \frac{1}{12}$.

Thus $100p = 8\frac{1}{3}\,\%$ is the estimated proportion of defective items in the whole bulk.

69. Exercises.

1. Razor blades of a certain kind are sold in packets of five. The following table shows the frequency distribution of 100 packets according to the number of faulty blades contained in them:

No. of faulty blades	0	1	2	3	4	5
No. of packets	80	17	2	1	0	0

Calculate the mean number of faulty blades per packet and, assuming that the binomial law applies, estimate the probability that a blade taken at random from any packet will be faulty. [Northern]

2. The production of an electrical component is checked by examining samples of 6. The following table shows the frequency distribution of 40 samples according to the number of defective components contained in each:

No. of defective components in sample of 6	0	1	2	3	4	5 or 6
No. of samples	12	16	8	3	1	0

Estimate the proportion of defective components in the whole output.

6

THE POISSON DISTRIBUTION

70. Introductory. Consider the four sampling-bottles shown in fig. 10. Bottle A contains a large number of steel balls of which 10 % are black and random samples of 5 are drawn from it. The probabilities $P(0)$, $P(1)$, $P(2)$, $P(x \geqslant 3)$ of 0, 1, 2, 3 *or more* black balls per sample are calculated, as shown in §60, by the binomial expansion of $(q+p)^n$ where $p = \frac{1}{10}$, $q = \frac{9}{10}$ and $n = 5$. Bottle B contains a large number of steel balls of which 5 % are black and random samples of 10 are drawn from it. Thus, for bottle B, $p = \frac{1}{20}$, $q = \frac{19}{20}$ and $n = 10$. Similarly, for bottle C, $p = \frac{1}{40}$, $q = \frac{39}{40}$ and $n = 20$ while for bottle D, $p = \frac{1}{100}$, $q = \frac{99}{100}$ and $n = 50$. In bottles A, B, C and D, therefore, p decreases and n increases in such a way that if a large number of samples are drawn from any bottle the mean number of black balls per sample, np, will be the same for all four bottles. Table 6A gives the probabilities $P(0)$, $P(1)$, $P(2)$, $P(x \geqslant 3)$ for each of the four bottles.

TABLE 6A

The binomial distribution of probabilities

No. of black balls per sample x	Bottle A $p = \frac{1}{10}$ $n = 5$ $P(x)$	Bottle B $p = \frac{1}{20}$ $n = 10$ $P(x)$	Bottle C $p = \frac{1}{40}$ $n = 20$ $P(x)$	Bottle D $p = \frac{1}{100}$ $n = 50$ $P(x)$
0	$P(0) = 0.590$	$P(0) = 0.598$	$P(0) = 0.603$	$P(0) = 0.603$
1	$P(1) = 0.328$	$P(1) = 0.315$	$P(1) = 0.309$	$P(1) = 0.304$
2	$P(2) = 0.073$	$P(2) = 0.075$	$P(2) = 0.075$	$P(2) = 0.075$
3 or more	$P(x \geqslant 3) = 0.009$	$P(x \geqslant 3) = 0.012$	$P(x \geqslant 3) = 0.013$	$P(x \geqslant 3) = 0.018$

When p is small and n is large as in the case of bottle D it is more convenient to calculate $P(0)$, $P(1)$, $P(2)$, ... from the terms of

$$e^{-a} \times \{\text{expansion of } e^a\}$$

$$= e^{-a}\left\{1+a+\frac{a^2}{2!}+\dots\right\}$$

$$= e^{-a}+ae^{-a}+\frac{a^2}{2!}e^{-a}+\dots,$$

where a is the mean number of black balls per sample.

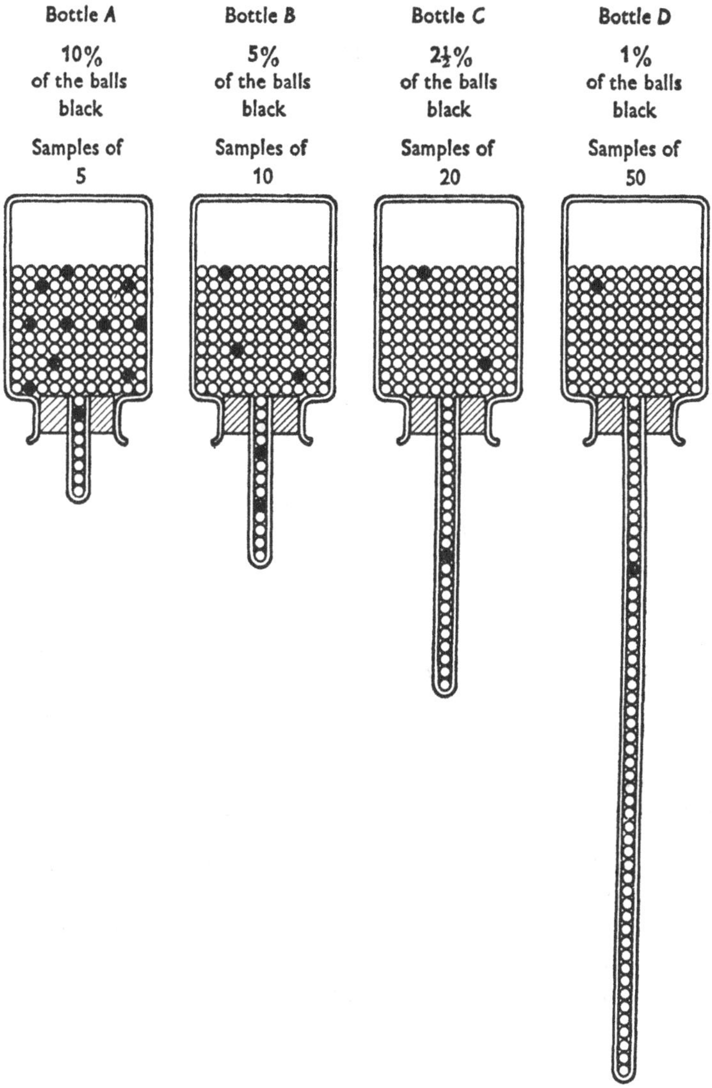

Fig. 10. Four sampling-bottles in which the proportion, p, of black balls decreases while the number, n, in a sample increases so that the mean number of black balls per sample, np, remains constant.

Thus a replaces np and in the example under consideration $a = \frac{1}{2}$.

The set of probabilities

$$e^{-a},\ ae^{-a},\ \frac{a^2}{2!}e^{-a},\ \frac{a^3}{3!}e^{-a},\ \ldots$$

is known as the Poisson distribution. Their formal derivation from the binomial distribution is given in §72. They have been used to calculate the probabilities of table 6B which, it will be seen, approximate very closely indeed to the probabilities for bottle *D* in table 6A and quite closely to the probabilities for bottles *A*, *B* and *C*.

TABLE 6B

The Poisson distribution of probabilities

$$e^{-a},\quad ae^{-a},\quad \frac{a^2}{2!}e^{-a},\quad \ldots,\quad \textit{where } a = \tfrac{1}{2}$$

No. of black balls per sample x	Probability $P(x)$
0	$P(0) = e^{-\frac{1}{2}} = 0.607$
1	$P(1) = \frac{1}{2}e^{-\frac{1}{2}} = 0.303$
2	$P(2) = \frac{(\frac{1}{2})^2}{2!}e^{-\frac{1}{2}} = 0.076$
3 or more	$P(x \geqslant 3) = 1-\{P(0)+P(1)+P(2)\} = 0.014$

71. Exercise.

One percent of the articles produced in a certain workshop are defective. Calculate the probabilities $P(0)$, $P(1)$, $P(2)$, $P(3)$, $P(4)$, $P(x \geqslant 5)$ of 0, 1, 2, 3, 4, 5 or more defective articles in a random sample of 200,

(i) by substituting $p = 0.01$, $q = 0.99$ and $n = 200$ in the binomial distribution,

(ii) by substituting $a = 2$ in the Poisson distribution.

(Make use of the values of e^{-x} given on page 201.)

72. The derivation of the Poisson distribution. In the formal consideration of the binomial and Poisson distributions it is usual to call p the probability of *success* and q $(= 1-p)$ the probability of *failure*. The Poisson distribution is the form assumed by the binomial distribution when p is small and n is large, the mean number of successes np being a finite constant a.

In $(q+p)^n$ let $p = a/n$. The probabilities of 0, 1, 2, 3, ... successes

$$q^n,\quad nq^{n-1}p,\quad \frac{n(n-1)}{2!}q^{n-2}p^2,\quad \frac{n(n-1)(n-2)}{3!}q^{n-3}p^3,\quad \ldots$$

then become

$$q^n,\quad aq^{n-1},\quad \frac{1(1-1/n)}{2!}a^2q^{n-2},\quad \frac{1(1-1/n)(1-2/n)}{3!}a^3q^{n-3},\quad \ldots.$$

When n is large, p is small, q is approximately unity and the terms of the distribution are proportional to

$$1,\quad a,\quad \frac{a^2}{2!},\quad \frac{a^3}{3!},\quad \ldots,$$

whose sum is e^a. The sum of the actual probabilities, however, is unity since it is the limit of $(p+q)^n$. The actual probabilities are therefore obtained by dividing each of the above terms by their sum e^a. Thus the probabilities of 0, 1, 2, 3, ... successes are

$$e^{-a}, \quad ae^{-a}, \quad \frac{a^2}{2!}e^{-a}, \quad \frac{a^3}{3!}e^{-a}, \quad \ldots$$

respectively.

73. The mean and variance of the Poisson distribution. By setting down the Poisson distribution as shown in table 6C its mean and variance can easily be deduced.

TABLE 6C

(i) No. of successes x	(ii) Probability or relative frequency f	(iii) fx (i) × (ii)	(iv) fx^2 (i) × (iii)
0	e^{-a}	0	0
1	ae^{-a}	ae^{-a}	ae^{-a}
2	$\frac{a^2}{1.2}e^{-a}$	$\frac{a^2}{1}e^{-a}$	$\frac{2a^2}{1}e^{-a}$
3	$\frac{a^3}{1.2.3}e^{-a}$	$\frac{a^3}{1.2}e^{-a}$	$\frac{3a^3}{1.2}e^{-a}$
⋮	⋮	⋮	⋮
Total	Σf	Σfx	Σfx^2

Now
$$\Sigma f = 1,$$
$$\Sigma fx = ae^{-a}\left\{1+\frac{a}{1}+\frac{a^2}{1.2}+\ldots\right\}$$
$$= ae^{-a}e^a$$
$$= a,$$
and
$$\Sigma fx^2 = ae^{-a}\left\{1+\frac{2a}{1}+\frac{3a^2}{1.2}+\ldots\right\}$$
$$= ae^{-a}\left\{\left(1+\frac{a}{1}+\frac{a^2}{1.2}+\ldots\right)+\left(\frac{a}{1}+\frac{2a^2}{1.2}+\ldots\right)\right\}$$
$$= ae^{-a}\{e^a+ae^a\}$$
$$= a+a^2.$$

Hence the mean,
$$\frac{\Sigma fx}{\Sigma f} = a$$
and the variance,
$$\frac{\Sigma fx^2}{\Sigma f} - \left(\frac{\Sigma fx}{\Sigma f}\right)^2 = a.$$

The variance of a Poisson distribution is, therefore, equal to the mean of the distribution.

74. An example in which p is small and n is large. *Before a consignment of potatoes may be exported as seed potatoes a random sample of* 300 *has to be inspected and found free of infection. Show that if the average rate of infection in the consignment is* 1 %, *then there is approximately a* 95 % *chance that it will not be passed.* [Cambridge]

The mean rate of infection for a sample of 300 is 3 and substituting $a = 3$ in the Poisson formula we find that the probability of a sample being free from infection is

$$e^{-3} = 0.0498$$

$$= 5\,\% \text{ approximately.}$$

Hence the probability that the consignment will not be passed is approximately 95 %.

75. A traffic example. Let us imagine an open stretch of country road. We can assume that the vehicles pass freely along it, quite independent of each other and at completely random times. If 300 vehicles pass a certain point on the road in 2 h the average traffic rate is $2\frac{1}{2}$ vehicles per minute and the probabilities $P(0)$, $P(1)$, $P(2)$, ... of 0, 1, 2, ... vehicles passing the point in any minute can be obtained by substituting $a = 2\frac{1}{2}$ in the Poisson formula (see table 6D).

TABLE 6D

No. of vehicles in any minute x	Probability $P(x)$
0	$P(0) = e^{-2.5} = 0.082$
1	$P(1) = 2.5e^{-2.5} = 0.205$
2	$P(2) = \frac{2.5^2}{2!}e^{-2.5} = 0.256$
3	$P(3) = \frac{2.5^3}{3!}e^{-2.5} = 0.213$
4 or more	$P(x \geqslant 4) = 1-\{P(0)+P(1)+P(2)+P(3)\} = 0.244$

76. Exercises.

1. Certain mass-produced articles, of which 0.5 % are defective, are packed in cartons each containing 100. What proportion of cartons are free from defective articles and what proportion contain 2 or more defectives? [R.S.S.]

2. Assuming that breakdowns in a certain electric power supply occur according to the Poissonian law with an average of one breakdown in 10 weeks calculate the probabilities of 0, 1, 2 or more breakdowns in any period of one week.

3. If the average number of calls made on a certain telephone route is 20 per hour calculate the probabilities of 0, 1, 2, 3, 4 or more calls being made in any period of 3 min.

4. When 3 dice are cast, the probability of obtaining 3 aces is $\frac{1}{216}$. If the 3 dice are cast 108 times and a prize is awarded every time 3 aces are obtained, calculate the probabilities of 0, 1, 2 or more prizes being awarded.

5. In a large consignment of eggs the average number broken in a crate is 3. Estimate the probability of a crate containing (i) no broken eggs, (ii) more than 3 broken eggs.

6. Assuming that railway accidents occur with relative frequencies that conform to the Poissonian law and that there is on the average 1 accident in 2 years, estimate the probability of 2 accidents occurring within a period of 3 months.

7. If, on the average each year, 1 miner out of every 2000 loses his life in a colliery accident, use the Poisson distribution to estimate what chance a colliery which employs 800 miners has of being free of fatal accidents in any given year.

8. Show that for a Poisson distribution the variance equals the mean. Are the following results likely to have come from a Poisson distribution?

No. of defects per piece of cloth	0	1	2	3	4	5	Total
Frequency of pieces	15	26	21	19	8	3	92

[R.S.S.]

9. Compute the average m and standard deviation σ of the following distribution and verify that approximately $\sigma = \sqrt{m}$.

x	0	1	2	3	4	5	Total
f	20983	2615	183	14	2	3	23800

Assuming that the frequencies are given by the formula $Nm^r e^{-m}/r!$ where $N = 23800$ and r takes successive values 0, 1, 2, 3, 4, 5, compute the expected frequencies from this formula taking m equal to the computed average.

[R.S.S.]

10. Derive the Poisson distribution

$$e^{-m}\left(1+m+\frac{m^2}{1.2}+\ldots\right)$$

as a limiting form of the binomial distribution $(p+q)^n$. Find the mean and standard deviation for the table of deaths of women over 85 years old recorded in *The Times* in a three-year period.

No. of deaths recorded on the day	0	1	2	3	4	5	6	7
No. of days	364	376	218	89	33	13	2	1

Find the expected number of days with 'one death recorded' for the Poisson series fitted to the data. [R.S.S.]

11. A car-hire firm has 2 cars, which it hires out by the day. The number of demands for a car on each day is distributed as a Poisson distribution, with mean 1.5.

Calculate the proportion of days on which neither of the cars is used, and the proportion of days on which some demand is refused.

If each car is used an equal amount, on what proportion of days is a given one of the cars not in use? What proportion of demands has to be refused?

[R.S.S.]

7

THE χ^2-DISTRIBUTION

77. The null hypothesis. In many examples and exercises so far, we have *assumed* without question that a distribution is normal, binomial or Poisson. By means of Pearson's χ^2-distribution (pronounced *kye-squared* and often printed *chi-squared*) shown in table A4, page 199, it is possible to estimate the extent to which such an *assumption* or *null hypothesis* is justified. The χ^2-distribution has other uses, illustrations of which will be given in this and the next chapter. Its mathematical derivation, however, is completely beyond the scope of this volume.

78. The use of χ^2 for testing normality. The distribution of the speeds shown in table 7A is the result of observations made at a certain place on a motorway. A cursory inspection of the distribution, or its histogram, reveals that it is roughly normal. Thus, if we regard it as a sample of the traffic in general at that place we are led to expect that the speeds of the cars and light commercial vehicles are normally distributed. To decide more definitely whether the distribution of speeds is normal or not the first step is to use the mean, 49.4 m.p.h., and the standard deviation 7.22 m.p.h. to construct a normal distribution with the same group intervals, as shown in table 7B. The frequencies of table 7A are called the *observed* frequencies and those of table 7B the *expected* frequencies. The former are the definite *observations* of a statistical investigation. The latter are to be *expected* if the *assumption* or *null hypothesis* that the distribution is normal is true. The question now arises: *Are the differences between the observed and expected frequencies great enough to force us to reject the null hypothesis as false or are they small enough to allow us to accept it as true?* In statistics we generally put this question as: *Are the differences* SIGNIFICANT *or not?*

TABLE 7A

Observed frequencies

Speeds in m.p.h. of 256 *private cars and light commercial vehicles*

Speed (m.p.h.)	30–	35–	40–	45–	50–	55–	60–	65–	70–	75–80
Frequency	2	22	48	65	72	29	11	5	0	2

TABLE 7B

Expected frequencies

Normal distribution of speeds having the same mean, standard deviation and total as the speeds of table 7A

Speed (m.p.h.)	Less than 35	35–	40–	45–	50–	55–	60–	65–	70 and over
Frequency	6	19	45	67	63	38	14	3	1

79. The calculation of χ^2. To answer the question of the last paragraph it is necessary to calculate χ^2 by the formula

$$\chi^2 = \Sigma\left[\frac{(O-E)^2}{E}\right],$$

where O is the observed frequency of a particular class and E is the corresponding expected frequency.

TABLE 7C

Calculation of χ^2

Speed (m.p.h.)	Observed frequency O	Expected frequency E	$O-E$	$\frac{(O-E)^2}{E}$ to 2 decimals
Less than 40	24	25	−1	0.04
40–	48	45	+3	0.20
45–	65	67	−2	0.06
50–	72	63	+9	1.28
55–	29	38	−9	2,13
60 and over	18	18	0	0.00
Total	256	256	0	3.71 = χ^2

The actual calculation is shown in table 7C. One important point needs some explanation. The mathematical derivation of the χ^2-distribution requires that all the expected frequencies shall be *sufficiently large*. 'Sufficiently large' in this case is usually taken to mean *a minimum of* 5. Bearing this in mind the classes at the beginning and end of tables 7A and 7B have been combined before entering them into table 7C.

80. The number of degrees of freedom, ν. Having calculated $\chi^2 = 3.71$ we need to note the *number of degrees of freedom*, ν (pronounced *new*) that were available in its calculation. The concept of 'degrees of freedom' is not an easy one. The method of calculating ν in each of the examples which follow will be carefully described and this should enable the student to develop a working knowledge of it. In table 7C there are 6 pairs of O and E values or 6 *classes*. The *number of restrictions* imposed in calculating the expected frequencies were 3 in that the expected frequencies have the same *total, mean* and *standard deviation* as the observed frequencies. This implies that if the E values were written down as 25, 45, 67, x, y, z the values of x, y, z could be calculated from 3 equations obtained from the 3 restrictions. Thus, although there are 6 classes the number of degrees of freedom available for the calculation of χ^2

$$\begin{aligned}\nu &= \text{the number of classes} - \text{the number of restrictions} \\ &= 6 \text{ classes} - 3 \text{ restrictions} \\ &= 3.\end{aligned}$$

Table A3, page 198, summarises the rules by which the student can obtain ν when testing the common types of distributions and contingency tables.

81. The acceptance or rejection of the null hypothesis. Table A4, page 199, gives the percentage points, P, of the χ^2-distribution for different values of ν. The column of values for $P = 5$ are of special importance and a few of these values are reprinted for convenience in table 7D.

TABLE 7D

The $P = 5\%$ values of the χ^2-distribution

ν	χ^2
1	3.84
2	5.99
3	7.81
4	9.49
5	11.07

If χ^2 is greater than the $P = 5\%$ value, the differences between the observed and expected frequencies are usually taken to be so significant that the null hypothesis must be rejected.

The value of χ^2, 3.71, for $\nu = 3$ is certainly not greater than the $P = 5\%$ value, 7.81. We conclude, therefore, that the differences are not significant and there are no grounds for rejecting the hypothesis. This means that we are justified in stating that the speeds of cars and light commercial vehicles are normally distributed about a mean 49.4 m.p.h. with a standard deviation of 7.22 m.p.h.

82. More precise interpretation of the P values. The various illustrations of the uses of the χ^2-distribution which follow will help to make clear the mathematical meaning of the P values. In the meantime table 7E is a descriptive treatment which will help in the interpretation of a value of χ^2 when it has been calculated. It indicates that, since the value of χ^2, 3.71, obtained in the last paragraph is considerably less than the $P = 10\%$ value, there is every reason to believe that the sample under consideration is from a normal population.

TABLE 7E

WHEN TO REJECT THE NULL HYPOTHESIS

If χ^2 is:	*The differences between the observed and expected frequencies are:*	*The null hypothesis is:*
Greater than $P = 5\%$ value	Significant	Probably false
Greater than $P = 2.5\%$ value	Very significant	Very probably false
Greater than $P = 1\%$ value	Most significant	Almost certainly false

WHEN TO ACCEPT THE NULL HYPOTHESIS

If χ^2 is:	*The differences between the observed and expected frequencies are:*	*The null hypothesis is:*
Less than $P = 5\%$ value	Not significant	Probably true
Less than $P = 10\%$ value	Not in the least significant	Very probably true

WHEN χ^2 IS INCREDIBLY SMALL

If χ^2 is less than $P = 95\%$ value the agreement of the observed data with the null hypothesis is *almost too good to be true* and it leads us to suspect:

(i) the sample is not random but carefully selected;
or (ii) the sample consists of fictitious data and is not the result of a proper statistical investigation;
or (iii) the null hypothesis may have been constructed in a ridiculously complicated way so as to suit the data too perfectly.

TABLE 7F

Speeds in m.p.h. of 256 cars and light commercial vehicles together with 32 heavy commercial vehicles

Speed (m.p.h.)	20–	25–	30–	35–	40–	45–	50–	55–	60–	65–	70–	75–80
Frequency	3	14	14	24	49	65	72	29	11	5	0	2

83. A distribution which is not normal. Table 7F shows the distribution of the speeds of the previous 256 cars and light commercial vehicles together with the 32 heavy commercial vehicles which also passed the point of observation. This raises the total number of vehicles observed to 288, lowers the mean speed to 47.25 m.p.h. and extends the standard deviation to 9.23 m.p.h. An inspection of the distribution, or its histogram, reveals that it no longer seems normal. The calculation of $\chi^2 = 22.39$ is shown in table 7G. As the number of classes in this case is 9 and the number of restrictions is again 3, the number of degrees of freedom $\nu = 6$. Reference to table A4 shows that χ^2 is greater than the $P = 0.5\ \%$ value and is almost equal to the $P = 0.1\ \%$ value. The differences between the observed and expected frequencies are, therefore, most significant and the null hypothesis that the distribution is normal must be rejected.

Thus the inclusion of the speeds of the heavy vehicles destroys completely the normality of the distribution.

TABLE 7G

Calculation of χ^2

Speed (m.p.h.)	O	E	$O-E$	$\frac{(O-E)^2}{E}$ to 2 decimals
Less than 30	17	9	+8	7.11
30–	14	17	−3	0.53
35–	24	36	−12	4.00
40–	49	54	−5	0.46
45–	65	62	+3	0.15
50–	72	52	+20	7.70
55–	29	34	−5	0.75
60–	11	16	−5	1.56
65–	7	8	−1	0.13
Total	288	288	0	$22.39 = \chi^2$

The fact that χ^2 is greater than the $P = 0.5\ \%$ value means that the probability of obtaining a random sample with a distribution like that of table 7F from a normal population is less than $0.5/100 = \frac{1}{200}$. Since χ^2 is almost equal to the $P = 0.1\ \%$ value this last probability is almost $0.1/100 = \frac{1}{1000}$. This precise mathematical interpretation of the value of χ^2 gives definite meaning to the rather vague statement 'almost certainly false' of table 7E.

84. Exercises.

Test each of the following distributions for normality.

1. In estimating the valuation of a plantation of fir trees, the girths of the trees in a sample area of 500 trees were measured and tabulated with a 0·2 m grouping interval as follows:

Girth (m)	0·4–0·6	0·6–0·8	0·8–1·0	1·0–1·2	1·2–1·4	1·4–1·6	1·6–1·8
No. of trees	25	30	135	160	100	40	10

[London]

2. *Chest girth of* 1000 *men (British) aged* 20 *in* 1939

Chest girth (mm)	Frequency
Under 800	26
800–	116
850–	304
900–	352
950–	161
1000–	34
1050 and over	7
Total	1000

[SOURCE: Martin, *The Physique of Young Adult Males*]

3. *Height of* 1000 *men (British) aged* 20

Height (mm)	Frequency
Under 1550	8
1550–	33
1600–	122
1650–	254
1700–	289
1750–	196
1800–	74
1850 and over	24
Total	1000

[Same source as Ex. 2 above]

4. *The frequency distribution of the weights of* 1000 *packets of tea delivered by an automatic packing machine*

Weight of packet of tea (g)	112–	114–	116–	118–	120–	122–124
Frequency	14	56	347	423	140	20

85. Testing a binomial distribution. Table 7H shows the *observed* and *expected* frequencies for an experiment in which 6 dice were thrown together 216 times. The expected frequencies are based on the assumption or null hypothesis that the distribution is binomial and the only restriction imposed in their calculation is that their total is made the same as that of the observed frequencies. The number of degrees of freedom available in calculating χ^2 is, therefore,

$$\nu = 4 \text{ classes} - 1 \text{ restriction}$$
$$= 3.$$

Since the value χ^2, 5.84, *is less than the $P = 10\,\%$ value* 6.25 *obtained from table* A4, *the null hypothesis is very probably true and it is safe to say that the distribution is binomial.*

TABLE 7H

Testing a binomial distribution

No. of sixes obtained when 6 dice are cast	Observed frequency (the result of an actual experiment) O	Expected frequency (based on the assumption or null hypothesis that the distribution is binomial) E	$\frac{(O-E)^2}{E}$
0	77	$216(\frac{5}{6})^6 = 72$	0.35
1	84	$216.6(\frac{5}{6})^5(\frac{1}{6}) = 87$	0.10
2	34	$216.15(\frac{5}{6})^4(\frac{1}{6})^2 = 43$	1.89
3 or more	21	$216 - \text{above} = 14$	3.50
Total	216	216	$5.84 = \chi^2$

86. Exercises.

Use the χ^2 test to determine if the following distributions conform to the binomial law. (Note that in Exs. 1 and 2 the *means* and *totals* agree and, therefore, the number of restrictions is 2. In Exs. 3, 4, 5 only the totals agree and the number of restrictions is 1 as in §85.)

1. The distribution of §69, Ex. 2.

2. The distribution of §68.

3.

No. of sixes obtained when 8 dice are cast	0	1	2	3 or more	Total
Frequency	16	22	14	8	60

4.

No. of fives or sixes when 6 dice are cast	0	1	2	3	4 or more	Total
Frequency	16	36	32	28	8	120

5.

No. of heads when 5 coins are tossed	0	1	2	3	4	5	Total
Frequency	25	123	246	264	146	28	832

Exercises 3–5 above are the results of experiments carried out by the author. The student will find it amusing to carry out similar experiments himself.

87. Testing a Poisson distribution. *A company which manufactures tubes for television receivers conducted a test of a sample batch of* 1000 *tubes and recorded the number of faults in each tube in the following table*:

No. of faults	0	1	2	3	4	5	6
Frequency	620	260	88	20	8	2	2

[London]

As the variance, 0.74, is approximately equal to the mean, $a = 0.55$, one might suspect that the distribution is Poissonian. Table 7I shows the calculation of χ^2. As the expected frequency of 4 or more faults is considerably less than 10 the last two classes have been combined.

TABLE 7I

Testing a Poisson distribution

No. of faults	Observed frequency (results of the actual test) O	Expected frequency (based on the assumption or null hypothesis that distribution is Poissonian) E	$\frac{(O-E)^2}{E}$
0	620	$1000 \times e^{-a} = 577$	3.20
1	260	$1000 \times ae^{-a} = 317$	10.25
2	88	$1000 \times \frac{1}{2}a^2e^{-a} = 87$	0.01
3	20 }	$1000 \times \frac{1}{6}a^3e^{-a} = 16$ }	8.90
4 or more	12 }	$1000 - \text{above} = 3$ }	
Total	1000	1000	$22.36 = \chi^2$

In calculating the expected frequencies two restrictions have been imposed, in that their mean and total have been made equal respectively to the mean and total of the observed frequencies. Hence the number of degrees of freedom, ν, available in the calculation of χ^2, is

$$4 \text{ classes} - 2 \text{ restrictions} = 2.$$

On reference to table A4 we find $\chi^2 = 22.36$ is greater than the $P = 0.1\%$ value and we infer that the observed frequencies differ very significantly from the expected frequencies. Therefore the null hypothesis that the distribution is Poissonian must be rejected.

88. A distribution may conform to both the binomial and the Poisson laws. As the Poisson distribution is a convenient alternative form of the binomial distribution when p is small and n is large many distributions will be found to conform to both laws. For example, the distribution of §85 (in which p has the comparatively small value of $\frac{1}{6}$ and n the comparatively large value of 6) does not differ significantly from the Poisson distribution with $a = 1$. Some of the exercises which follow are further illustrations of this point.

89. Exercises.

Use the χ^2-test to determine if the following distributions conform to the Poisson law.

1. The distribution of §76, Ex. 8.
2. The distribution of §76, Ex. 9.
3. The distribution of §76, Ex. 10.
4. The distribution of §69, Ex. 2.
5. The distribution of §68.
6. The distribution of §86, Ex. 3.
7. The distribution of §86, Ex. 4.
8. The distribution of §86, Ex. 5.

8

THE USE OF χ^2 IN TESTING CONTINGENCY TABLES

90. Introductory experiment. Throwing a single die. The following table gives the frequency distribution obtained when a single die was cast 648 times.

Score	1	2	3	4	5	6
Frequency	96	98	117	130	107	100

If the die were unbiased the expected frequency of each score would be 108 since each number on the die would be equally probable. The χ^2-test can be used to decide if the observed frequencies differ significantly from the expected frequencies. The calculation of χ^2 is shown in table 8A.

TABLE 8A

Testing a die

Score	Observed frequencies (the result of an actual experiment) O	Expected frequencies (based on the assumption or null hypothesis that each score is equally probable) E	$\frac{(O-E)^2}{E}$
1	96	108	1.33
2	98	108	0.93
3	117	108	0.75
4	130	108	4.48
5	107	108	0.01
6	100	108	0.59
Total	648	648	$8.09 = \chi^2$

No. of degrees of freedom, ν = 6 classes − 1 restriction (totals equal)
= 5.

As the value of χ^2, 8.09, is less than the $P = 10\,\%$ value, 9.24, we conclude that the differences between the observed and expected frequencies are not significant and we accept the null hypothesis that each score was equally probable in spite of the high frequency of the score 4.

91. An experiment with playing cards. After an ordinary pack of 52 playing cards comprising 4 suits of 13 spades, 13 hearts, 13 diamonds and 13 clubs had been well shuffled, a single card was drawn at random from it and the suit noted. This experiment was repeated 98 times and the following distribution obtained.

Suit	Spades	Hearts	Diamonds	Clubs
Frequency	23	31	20	24

The expected frequencies in this case were each $24\frac{1}{2}$ because the suits were equally probable for each random selection. The question arose: 'Had the differences in the frequencies some special significance or were they merely the fluctuations of experiment?' Table 8B shows the application of the χ^2-test.

TABLE 8B

Testing a pack of cards by selecting one card at random from the pack after it had been thoroughly shuffled

Suit	Observed frequencies O	Expected frequencies (based on the hypothesis that each suit was equally probable) E	$\frac{(O-E)^2}{E}$
Spades	23	$24\frac{1}{2}$	0.09
Hearts	31	$24\frac{1}{2}$	1.72
Diamonds	20	$24\frac{1}{2}$	0.83
Clubs	24	$24\frac{1}{2}$	0.01
Total	98	98	$2.65 = \chi^2$

No. of degrees of freedom, ν = 4 classes − 1 restriction (totals equal)
= 3.

As the value of χ^2, 2.65, is less than the $P = 10\%$ value, 6.25, the differences were not significant. They could be accepted as the normal fluctuations of experiment.

92. Contingency tables. When the classes in which frequencies are grouped are not class-intervals in a measured variate, but correspond to some attribute or descriptive quality the frequency table is called a *contingency table*. Section 91 provides a good introductory example of a contingency table. Other examples follow.

93. An example from an engineering workshop. In a certain workshop fifty of each of four types of cutting tool are in regular use and a record was kept for a period of 3 months of the number of breakages of each type.

These are presented in table 8C together with the calculation of χ^2 based on the null hypothesis that, in spite of the observed variations in breakages, all four types of cutting tool are equally good.

TABLE 8C

Testing four different types of cutting tool.
Null hypothesis—all types are equally good

Type	1	2	3	4	Total
Observed no. broken, O	21	13	20	16	70
Expected no. broken, E	$17\frac{1}{2}$	$17\frac{1}{2}$	$17\frac{1}{2}$	$17\frac{1}{2}$	70
$(O-E)^2/E$	0.70	1.16	0.36	0.13	2.35
Observed no. NOT broken, O	29	37	30	34	130
Expected no. NOT broken, E	$32\frac{1}{2}$	$32\frac{1}{2}$	$32\frac{1}{2}$	$32\frac{1}{2}$	130
$(O-E)^2/E$	0.38	0.62	0.19	0.07	1.26

When the values 21, 13, 20 have been written down as the observed numbers broken for types 1, 2, 3 all the $(O-E)$ pairs used in table 8C can be obtained by making use of the total 70 broken. Thus, although $\chi^2 = 3.61$ is calculated from 8 values of $(O-E)^2/E$, it is based on only 3 degrees of freedom. As χ^2 is less than the $P = 10\%$ value, 6.25, the differences are not significant and the null hypothesis is not rejected. We must, therefore, accept the cutting tools as equally good in spite of the variations in breakages.

Note that, as the number of tools in use increases, the $(O-E)^2/E$ values for those NOT broken decrease and eventually can be omitted.

After reading §94 the student will realise that it is easy to extend the calculation of table 8C so as to cover the case when different numbers of each type of tool are in use.

94. Expected frequencies in a given ratio. The following experiment is interesting in that it produces a contingency table in which the expected frequencies are in the ratio 4:3:2:1. From an ordinary pack of 52 playing cards, 12 spades, 9 hearts, 6 diamonds and 3 clubs were taken to form a *special* little pack of 30 cards. After shuffling this *special* pack a card was drawn at random from it and its suit noted. This experiment was repeated 170 times. Table 8D presents the observed frequencies and the calculation of χ^2 based on the null hypothesis that the expected frequencies of spades, hearts, diamonds, clubs are in the ratio 12:9:6:3. As χ^2 is less than the $P = 10\%$ value, the differences are not significant and the null hypothesis is upheld.

TABLE 8D

Testing a pack of 30 *playing cards comprising* 12 *spades,* 9 *hearts,* 6 *diamonds,* 3 *clubs, by drawing at random one card from the pack after shuffling it*

Suit	Observed frequency (result of 'shuffling and drawing' 170 times) O	Expected frequency (based on the null hypothesis that the frequencies would be in the ratio 12:9:6:3) E	$\frac{(O-E)^2}{E}$
Spades	73	68	0.37
Hearts	49	51	0.08
Diamonds	32	34	0.12
Clubs	16	17	0.59
Total	170	170	$1.16 = \chi^2$

No. of degrees of freedom, ν = 4 classes − 1 restriction (totals equal) = 3.

95. An example from genetic theory. *Genetic theory states that children having one parent of blood-type M and the other of blood-type N will always be one of the three types M, MN, N and the proportions of these types will on the average be as* 1:2:1. *A report states 'of* 162 *children having one M parent and one N parent,* 28.4 % *were found to be of type M,* 42 % *of type MN and the remainder of type N. The low value of* χ^2 *demonstrates the truth of the genetic theory.' Calculate the value of* χ^2, *make the appropriate test of significance and comment on the conclusions quoted.* [R.S.S.]

The calculation, which is similar to that of §94, is shown in table 8E. The observed frequencies 28.4 % of 162, 42 % of 162, 29.6 % of 162 are 46, 68, and 48 respectively. As χ^2 is less than the $P = 10\,\%$ value, 4.62, the null hypothesis is accepted. A significance test does not provide certain proof or disproof. The low value of χ^2 does not demonstrate that the genetic theory is true. It demonstrates that the data does not provide sufficient evidence for rejecting the genetic theory.

96. An example illustrating Yates's correction when $\nu = 1$. *A door-to-door salesman's records show that* 25 % *of his calls are successful. After taking a course in salesmanship* 12 *of the first* 30 *calls he makes are successful. Is this definite proof that his technique is improved?*

The null hypothesis in this case is that the apparent improvement is a chance effect and that, in the long run, the man's proportion of successful calls will still be only 25 %. The calculation of χ^2 is shown in table 8F. As there are only two classes and the totals are equal the number of degrees

TABLE 8E

An example from genetic theory

Blood-type	Observed frequency O	Expected frequency (based on the null hypothesis that the proportions are 1:2:1) E	$\frac{(O-E)^2}{E}$
M	46	$40\frac{1}{2}$	0.75
MN	68	81	2.09
N	48	$40\frac{1}{2}$	1.39
Total	162	162	$4.23 = \chi^2$

No. of degrees of freedom, ν = 3 classes − 1 restriction (totals equal) = 2.

of freedom is unity and it is necessary to apply *Yates's correction for continuity* which states that *when $\nu = 1$ the $(O-E)$ differences must each be diminished numerically by $\frac{1}{2}$.*

The reason for using Yates's correction is that χ^2 is a continuous distribution and we are using it, for contingency tables, on discrete results. This is permissible if there are several $(O-E)$ values but not if there is only one.

TABLE 8F

Calculation of χ^2 using Yates's correction when $\nu = 1$

	Observed frequencies O	Expected frequencies (based on the null hypothesis that 25 % are successful) E	Yates's correction of $(O-E)$: $(O-E)$	Y	$\frac{Y^2}{E}$
Successes	12	$7\frac{1}{2}$	$4\frac{1}{2}$	4	2.14
Failures	18	$22\frac{1}{2}$	$-4\frac{1}{2}$	−4	0.71
Total	30	30	0	0	$2.85 = \chi^2$

As the value of χ^2 is less than the $P = 5\%$ value, the differences are not significant. The null hypothesis is, therefore, not rejected and we conclude that there is no definite proof that the salesman's technique is improved.

97. Exercises.

1. While practising rifle shooting a man is successful with $\frac{1}{3}$ of his shots. After special coaching he achieves 19 successes with his first 40 shots. Use χ^2 to show that his marksmanship cannot be considered definitely improved.

If he were to go on shooting and obtain 38 successes with his first 80 shots would he then have proved himself a better shot?

2. Ten years ago the numbers of male and female office workers in a certain city were in the ratio 3:2. A recent random sample of 500 office workers revealed that 280 were men and 220 were women. Is this definite proof that the percentage of women office workers has increased?

3. A football pools expert claims that, over a long period, 50 % of his forecasts have been correct. On a particular Saturday he has only 24 forecasts out of 64 correct. Does this indicate that his claim is probably false?

98. A 2×2 contingency table. The following table summarises the *infant mortality* and *overcrowding* figures of 100 districts.

TABLE 8G

		Overcrowding		
		High	Low	Total
Infant mortality	High	22	15	37
	Low	14	49	63
Total		36	64	100

Thus, there are 22 districts in which the overcrowding is high and the infant mortality high, 15 in which the overcrowding is low and the infant mortality is high and so on. It is an example of a 2×2 contingency table. The question arises 'Does association exist between the two variables or are they completely independent?' Alternatively one might ask, 'Is a high infant mortality figure generally associated with a high overcrowding figure?'

Let us suppose that the variables are completely independent. A probability table can then be drawn up as shown in table 8H:

TABLE 8H

		Overcrowding		
		High	Low	Probability
Infant mortality	High	$0.36 \times 0.37 = 0.1332$	$0.64 \times 0.37 = 0.2368$	0.37
	Low	$0.36 \times 0.63 = 0.2268$	$0.64 \times 0.63 = 0.4032$	0.63
Probability		0.36	0.64	1

The marginal probabilities in the above table are obtained by dividing the marginal totals of the previous table by the total frequency 100. The marginal probabilities indicate that if any district is chosen at random from the 100 districts, the probability that it is highly overcrowded is 0.36 and the probability that its infant mortality is high is 0.37. Hence, if these probabilities are independent, the probability (see §40) of any district chosen at random having high overcrowding and high infant mortality figures is the product 0.36×0.37. Thus the probabilities in the four central *cells* are obtained from the marginal probabilities in the manner indicated.

The probability table can finally be converted into the table of expected frequencies (table 8I) by multiplying the probability in each *cell* by the total frequency 100.

TABLE 8I

Table of expected frequencies

(*Based on the null hypothesis that infant mortality and overcrowding are independent*)

		Overcrowding		
		High	Low	Total
Infant mortality	High	13.32	23.68	37.00
	Low	22.68	40.32	63.00
Total		36.00	64.00	100.00

The calculation of χ^2 is shown in table 8J. The number of degrees of freedom in a 2×2 table is unity. This is due to the fact that when one of the four values in the central cells is known the other three values can be obtained from the marginal totals which are the same in the O and E tables. Since $\nu = 1$ it is necessary to apply Yates's correction.

TABLE 8J

	Observed frequencies O	Expected frequencies E	$(O-E)$	Yates's correction of $(O-E)$ Y	$\frac{Y^2}{E}$
	22	13.32	8.68	8.18	5.02
	15	23.68	−8.68	−8.18	2.83
	14	22.68	−9.68	−8.18	2.95
	49	40.32	8.68	8.18	1.66
Total	100	100.00	0.00	0.00	$12.46 = \chi^2$

As the value of χ^2 is greater than the $P = 0.1\%$ value the null hypothesis that the two variables are independent is rejected. Association between infant mortality and overcrowding is, therefore, very probable.

99. Experiments using coloured dice.

1. *Throwing two dice.*

TABLE 8K

		Black die score	
		1–3	4–6
White die score	1–3	23	27
	4–6	21	29

Table 8K is a 2×2 contingency table in which association should not exist. It was obtained by throwing together a black die and a white die and one would expect the score of the white die to be completely independent of that of the black die. Draw up a table of expected frequencies based on the null hypothesis that no association exists between the pairs of scores and show by calculating χ^2 that the null hypothesis is probably true.

Perform a similar experiment for yourself.

2. *Throwing three dice.*

TABLE 8L

		Score of black and white dice	
		2–6	7–12
Score of black and green dice	2–6	26	16
	7–12	14	44

Table 8L is a 2×2 contingency table in which association should exist. It was obtained by throwing together a black die, a white die and a green die and as the black score is common to both totals the pairs are not completely independent. Test the table in a similar way to experiment 1.

Perform a similar experiment for yourself.

100. Exercises.

1. *Place of midday meal of earners, Stepney,* 1946

Earners with weekly fares	At home	Elsewhere	Totals
2*s*. 9*d*. and under	355	120	475
Over 2*s*. 9*d*.	19	85	104
Total	374	205	579

[Source: unpublished material from a sample survey of Stepney]

Assume that the sample is a random one from a large population. Is it possible that there was no association between the place of midday meal and fares in Stepney? [London]

2. A certain type of surgical operation can be performed either with a local anaesthetic or with a general anaesthetic. Results are as given below:

	Alive	Dead
Local	511	24
General	173	21

Test for any difference in the mortality rates associated with the different types of anaesthetic. [R.S.S.]

3. The following information was obtained in a sample of 50 small general shops:

	Shops in: Urban districts	Shops in: Rural districts	Total
Owned by men	17	18	35
Owned by women	3	12	15
Total	20	30	50

Can it be said that there are relatively more women owners of small general shops in rural than in urban districts? [London]

4. *Cigarette smoking and lung cancer*

	Death due to: Lung cancer	Death due to: Other causes
Heavy smokers	27	11
Light smokers	18	44

Do the above figures suggest that heavy smokers are more likely to die from cancer of the lung than light smokers?

(NOTE. This exercise is set as an example of the technique. The figures are not from any official source.)

5. *Greenwood and Yule's data on inoculation against typhoid*

	Attacked	Not attacked	Total
Inoculated	56	6759	6815
Not inoculated	272	11396	11668
Total	328	18155	18483

Do the above figures indicate that inoculation is a good preventative against typhoid?

101. The $h \times k$ contingency table. The method used for the 2×2 contingency table can be extended for use with a table which has h rows and k columns where h and k are both greater than 1. Since the marginal totals of E's and O's are made to agree the number of degrees of freedom in this case is

$$\nu = (h-1)(k-1),$$

because the last row and last column are determined once the remaining cells have been filled.

Table 8M shows the application of the χ^2-test to a 3×3 classification of the examination results in Pure Mathematics and Applied Mathematics of 200 candidates.

TABLE 8M

(i) The observed frequencies

		Pure Mathematics			
		Good	Pass	Fail	Total
Applied Mathematics	Good	16	13	10	39
	Pass	20	33	16	69
	Fail	10	28	54	92
Total		46	74	80	200

(ii) The expected frequencies (based on the assumption of no association)

		Pure Mathematics			
		Good	Pass	Fail	Total
Applied Mathematics	Good	$\frac{39}{200}\times\frac{46}{200}\times 200$	$\frac{39}{200}\times\frac{74}{200}\times 200$	$\frac{39}{200}\times\frac{80}{200}\times 200$	$\frac{39}{200}\times 200$
	Pass	$\frac{69}{200}\times\frac{46}{200}\times 200$	$\frac{69}{200}\times\frac{74}{200}\times 200$	$\frac{69}{200}\times\frac{80}{200}\times 200$	$\frac{69}{200}\times 200$
	Fail	$\frac{92}{200}\times\frac{46}{200}\times 200$	$\frac{92}{200}\times\frac{74}{200}\times 200$	$\frac{92}{200}\times\frac{80}{200}\times 200$	$\frac{92}{200}\times 200$
Total		$\frac{46}{200}\times 200$	$\frac{74}{200}\times 200$	$\frac{80}{200}\times 200$	200

(iii) The calculation of χ^2

										Total
O	16	13	10	20	33	16	10	28	54	200
E	8.97	14.43	15.60	15.87	25.53	27.60	21.16	34.04	36.80	200
$(O-E)^2/E$	5.51	0.14	2.01	1.07	2.19	4.88	5.89	1.07	8.04	$30.80 = \chi^2$

(iv) The number of degrees of freedom $h = k = 3$

$$\nu = (h-1)(k-1)$$
$$= 4.$$

As the value of χ^2 is greater than the $P = 0.1\,\%$ value, 18.47, the differences between the O's and E's are highly significant and the assumption of no association must be rejected. Therefore, one concludes, as might be expected, that association almost certainly exists between a candidate's results in Pure Mathematics and Applied Mathematics.

102. Exercises.

1. *Throwing two dice*

		Black die score		
		1 or 2	3 or 4	5 or 6
White die score	1 or 2	12	14	11
	3 or 4	17	9	9
	5 or 6	7	9	12

Show, by calculating χ^2, that probably no association exists between the pairs of scores in the above 3×3 table.

Perform a similar experiment for yourself.

2. *Relationship between number of wage-earners and output per manshift at coal mines employing* 100 *or more wage-earners in Great Britain in* 1945

Size of mine	No. of mines with an output per manshift of				
No. of wage-earners	Under 15 cwt	15 cwt and under 20 cwt	20 cwt and under 25 cwt	25 cwt and over	Total mines
100–499	103	140	76	42	361
500–999	58	131	76	39	304
1000 and over	25	73	83	48	229
Total mines	186	344	235	129	894

(Source: Ministry of Fuel and Power, *Statistical Digest*, 1945)

By calculating χ^2 show that there is probably association between the number of wage-earners employed in a mine and the output per manshift in that mine. [London]

3. *Brownlee's data on severity of smallpox attack*

Years since vaccination took place	Severity of attack				
	Very severe	Severe	Moderate	Light	Total
0–25	43	120	176	148	487
25–45	184	299	268	181	932
Over 45 or unvaccinated	111	89	40	30	270
Total	338	508	484	359	1689

Calculate χ^2 for the above 3×4 table and use it to demonstrate that association probably exists between severity of smallpox attack and the years that have elapsed since vaccination.

4. The first 10000 decimal places of $(\pi - 3)$ have the digits occurring with the following frequencies:

Digit	0	1	2	3	4	5	6	7	8	9
Frequency	968	1026	1021	974	1012	1046	1021	970	948	1014

[Source: *Mathematical Gazette*, No. 352]

Examine the hypothesis that there is no tendency for one digit to occur more often than another.

The frequencies of the respective digits for the thousand places 8001 to 9000 are 101, 103, 100, 103, 101, 99, 98, 97, 90, 108. Is there anything remarkable about these figures? [A.E.B.]

5. Four coins were tossed 4096 times, and the number of heads obtained was noted each time, with the following results:

Number of heads	0	1	2	3	4
Observed frequency	249	1000	1552	1050	245

Examine the hypothesis that the coins are unbiased. [A.E.B.]

6. At a public library, during a given week, the following numbers of books were borrowed:

Day	Mon.	Tue.	Wed.	Thur.	Fri.	Sat.
Number issued	204	292	242	283	252	275

Is there reason to believe that more books are generally borrowed on one weekday than on another? [A.E.B.]

7. In three firms, the number of employees in three categories are as follows:

Firm	*A*	*B*	*C*
Skilled manual	28	42	50
Unskilled	33	76	71
Non-manual	39	82	79

Apply a suitable test to find whether there is significant difference between the proportions of the various kinds of workers and state your conclusions. [A.E.B.]

8. Three schools, *A*, *B* and *C*, entered candidates for the same examination with the following result:

School	Pass	Fail
A	48	25
B	80	45
C	27	12

Use the χ^2 test to find whether there is any significant difference between the groups of candidates entered.

Comment upon the fairness of comparing schools by the percentage of candidates who pass examinations. [A.E.B.]

9. Pupils from three schools *A*, *B* and *C*, sit for an examination in Pure Mathematics and Applied Mathematics. The results are as follows:

	School			
	A	*B*	*C*	Total
Passed in both subjects	9	17	5	31
Passed in Pure Mathematics only	18	17	1	36
Passed in Applied Mathematics only	5	9	10	24
Failed in both subjects	6	2	1	9
Total	38	45	17	100

Examine the hypotheses (i) that there is no essential difference between the schools in Pure Mathematics, (ii) that there is no association between passing in Pure Mathematics and passing in Applied Mathematics, taking all the candidates together. [A.E.B.]

10. The following fictitious data refer to hair and eye colours for 200 persons:

Eye colour	Hair colour: Dark	Medium	Fair	Total
Blue	3	42	30	75
Grey	2	18	5	25
Brown	45	40	15	100
Total	50	100	50	200

Use the χ^2 test to examine the hypothesis that there is no association between hair and eye colours. [A.E.B.]

11. Six hundred factories were inspected and classified into four groups according to the number of persons employed: (i) under 50, (ii) 51 to 150, (iii) 151 to 250, (iv) over 250. They were also classified according to the condition of the premises (*a*) good, (*b*) satisfactory, (*c*) unsatisfactory. The results are shown in the following table (hypothetical data):

	(i)	(ii)	(iii)	(iv)	Totals
(*a*)	85	130	50	60	325
(*b*)	90	80	20	45	235
(*c*)	25	15	0	0	40
Totals	200	225	70	105	600

Use the χ^2 test to find whether it is likely that there is an association between the number employed and the condition of the factory. [A.E.B.]

12. The following table (Source: Kelsall, R. K. quoted in the *Robbins Committee's Report on Higher Education*) shows the type of school attended by Home entrants to Oxford and Cambridge, and to other universities in England and Wales in 1955.

	Percentage of entrants from various types of schools: Maintained	Direct grant	Independent	Numbers
Men				
Oxford and Cambridge	31	12	57	3709
Other universities	72	10	18	9647
Women				
Oxford and Cambridge	39	21	40	533
Other universities	66	14	20	4638

Estimate the numbers of entrants in the various groups and report on the significance or otherwise of the difference between the entrants to Oxford and Cambridge and to the other universities (*a*) among men, (*b*) among women.

[A.E.B.]

9

SAMPLES AND SIGNIFICANCE

103. Sample estimates of population values. To revise the use of a *random sample* to estimate the properties of the *parent population* from which the sample is drawn, consider the following example.

The following observations are heights in cm of a random sample of twenty compression springs taken from a current production in a works:

1.010	1.002	1.009	1.005	1.006
1.002	1.007	1.007	1.011	1.002
1.007	1.008	1.003	1.002	1.001
1.007	1.005	1.000	1.008	1.008

Calculate the mean and standard deviation of these observations and hence estimate the mean and standard deviation of the complete population of springs produced in the works.

If the distribution is normal, what is the probability that the height of any spring selected at random is

(i) *more than* 1.010 *cm;*

(ii) *more than* 1.015 *cm;*

(iii) *more than* 0.010 *cm from the mean?*

First rewrite the heights in thousandths with 1.005 cm as origin, thus:

5	−3	4	0	1
−3	2	2	6	−3
2	3	−2	−3	−4
2	0	−5	3	3

The estimate of the population mean, Est (μ), is then

$$m = \tfrac{1}{20}\Sigma x$$

$$= \tfrac{1}{20}\{5-3+4+\ldots+3\}$$

$$= 0.5 \text{ thousandths with } 1.005 \text{ cm as origin}$$

$$= 1.0055 \text{ cm},$$

and the estimate of the population standard deviation, Est (σ), is

$$s = \sqrt{\left\{\frac{\Sigma x^2}{19} - \frac{20}{19}\left(\frac{\Sigma x}{20}\right)^2\right\}}$$

$$= \sqrt{\left\{\frac{202}{19} - \frac{20}{19}(0.5)^2\right\}}$$

$$= 3.22 \text{ thousandths}$$

$$= 0.00322 \text{ cm.}$$

To answer question (i) it should be noted that a spring of height 1.010 cm has a *standardized deviate* from the mean of

$$\frac{1.010-1.0055}{0.00322} = 1.397$$

and by table A2, $A(1.397) = 0.9188$. Thus the probability of any spring being of greater height than 1.010 cm is

$$1-A(1.397) = 0.0812$$

$$= \tfrac{1}{12} \text{ approximately.}$$

Similarly to answer question (ii), a spring of height 1.015 cm has a standardized deviate of

$$\frac{1.015-1.0055}{0.00322} = 2.95$$

and $A(2.95) = 0.9984$. Hence the probability of a spring being of greater height than 1.015 cm is 0.0016 or $\frac{1}{625}$.

Finally to answer question (iii) it should be noted that a height which 0.010 cm from the mean has a standardized deviate of

$$\frac{0.010}{0.00322} = 3.106.$$

Since $A(3.106) = 0.99905$, the probability that any spring has a height of more than 0.010 cm *above* the mean is 0.00095. Further, the probability that any spring has a height of more than 0.010 cm *below* the mean is also 0.00095 and hence the *total* probability that any spring has a height more than 0.010 cm from the mean is 0.00190 or approximately $\frac{1}{500}$.

104. Two-tail probability. Question (iii) above provides an example of a *two-tail* probability. The answers to questions (i), (ii) and (iii) are illustrated diagrammatically by fig. 11 in which the total area under the graph is supposed to be unity and the areas *JKB*, *LMB* and *PQB* are supposed to be 0.0812, 0.0016 and 0.00095 respectively although they are not actually drawn to scale.

105. The frequency distribution of means of samples. If observations are collected, not individually, but as random samples of n, a frequency distribution can be constructed for the means of the samples. It is customary to represent the mean and standard deviation of this frequency distribution of sample-means as μ_n and σ_n respectively. Now it can be shown theoretically that

$$\mu_n = \mu,$$
$$\sigma_n = \sigma/\sqrt{n},$$

where μ and σ are the mean and standard deviation of the individual observations.

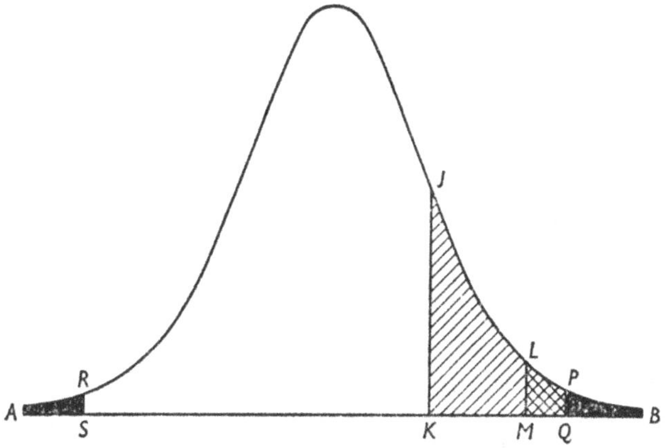

Fig. 11. Illustration of probabilities (i), (ii) and (iii) of §103 (not to scale). (i) Area *JKB* is the probability of the height of any spring being more than 1.010 cm. (ii) Area *LMB* is the probability of the height of any spring being more than 1.015 cm. (iii) The sum of the areas of the two tails *ARS* and *PQB* is the probability of the height of any spring being more than 0.010 cm from the mean.

A numerical illustration of this important theorem is provided by table 9A which gives the times of 256 servicings of a certain type of aircraft. These times, which vary from just over 2 hours to just under 4 hours, are divided into 32 groups of 8, the mean time of each group being given in the final column, to 1 decimal place.

The mean and standard deviation of the 256 individual times are

$$\mu = 168.4 \text{ m.p.h.},$$
$$\sigma = 14.66 \text{ m.p.h.}$$

The mean and standard deviation of the 32 sample-means are

$$\mu_8 = 168.4 \text{ m.p.h.},$$
$$\sigma_8 = 5.40 \text{ m.p.h.}$$

But
$$\sigma/\sqrt{8} = 14.66/2.828 \text{ m.p.h.},$$
$$= 5.19 \text{ m.p.h.}$$

TABLE 9A

Time, in minutes, for the servicing of a certain type of aircraft. 256 services studied in groups of 8

	Times, in minutes, for the 8 services of the group								Group mean time
Group 1	148	168	168	174	172	185	174	174	170.4
Group 2	160	171	178	157	178	183	168	168	170.4
Group 3	161	162	171	173	156	181	168	155	165.9
Group 4	171	168	176	169	196	175	157	171	172.9
Group 5	146	152	160	170	190	190	169	173	168.8
Group 6	145	169	165	174	199	182	178	170	172.8
Group 7	154	144	178	194	168	187	144	154	165.4
Group 8	163	165	168	171	184	172	169	184	172.0
Group 9	161	146	170	173	166	161	172	160	163.6
Group 10	147	160	161	153	174	187	152	150	160.5
Group 11	183	179	174	175	209	190	196	175	185.1
Group 12	173	164	178	169	209	164	154	145	169.5
Group 13	150	171	176	182	178	176	175	171	172.4
Group 14	148	168	176	178	183	186	170	159	171.0
Group 15	161	165	173	182	186	176	177	202	177.8
Group 16	143	169	174	171	181	167	168	146	164.9
Group 17	146	147	164	168	170	158	154	154	157.6
Group 18	156	164	178	148	182	158	186	154	165.8
Group 19	164	160	159	193	188	158	164	185	171.4
Group 20	158	163	164	179	178	172	177	163	169.0
Group 21	141	157	154	185	191	178	168	166	167.5
Group 22	152	154	189	176	177	189	187	189	176.6
Group 23	178	147	170	151	189	170	182	156	167.9
Group 24	161	150	160	178	173	175	161	162	165.0
Group 25	158	163	159	168	164	178	172	161	165.4
Group 26	156	181	203	208	169	190	151	155	176.6
Group 27	162	156	177	166	195	175	153	147	166.4
Group 28	150	140	157	184	220	169	140	171	166.4
Group 29	154	160	162	174	159	178	161	146	161.8
Group 30	150	136	149	185	150	224	164	152	163.8
Group 31	154	158	161	178	171	166	174	147	163.6
Group 32	152	156	173	167	159	163	170	136	159.5

Mean time of the 256 individual times = 168.4 minutes.
Standard deviation of the 256 individual times = 14.66 minutes.

Mean of the 32 group means = 168.4 minutes.
Standard deviation of the 32 group means = 5.40 minutes.

Thus table 9A gives an approximate verification of the general theorem for the case when $n = 8$.

106. Standard error and confidence limits of large samples. Suppose we have a single sample of n observations (n large). We may regard its mean m and standard deviation s as estimates of μ and σ, the mean and standard deviation of the parent population. Now $\sigma/\sqrt{n}$ is the standard deviation of the means of a large number of samples of n observations similar to our single sample. We can thus regard our single sample as one of many whose means are distributed with standard deviation $s/\sqrt{n}$ about a mean m, s being our estimate of σ. Moreover, it can be shown that even when the distribution of the whole population is far from normal the distribution of the sample-means will be approximately normal. Thus, in 95 % of cases m will lie in the range given by

$$\mu - 1.96\sigma/\sqrt{n} < m < \mu + 1.96\sigma/\sqrt{n}.$$

This gives $\quad \mu > m - 1.96\sigma/\sqrt{n} \quad$ and $\quad \mu < m + 1.96\sigma/\sqrt{n}.$

But we use s as our estimate of σ and hence *it is reasonably certain that the true value of the mean lies between*

$$m - 1.96s/\sqrt{n} \quad \text{and} \quad m + 1.96s/\sqrt{n}.$$

These limits are known as the 95 % *confidence limits* and $s/\sqrt{n}$ is called the *standard error of the mean.*

The probability of the true mean being outside the range $m \pm 1.96s/\sqrt{n}$ *is* 0.05 *or* $\frac{1}{20}$.

Further reference to table A2 shows that 99.8 % of the sample-means lie between $m - 3.09s/\sqrt{n}$ and $m + 3.09s/\sqrt{n}$. These, therefore, are the 99.8 % *confidence limits* and the probability of the true mean being outside them is 0.002 or $\frac{1}{500}$.

107. Use of the *t*-distribution for small samples. Consider Machine A, §9, Ex. 3, which provides a sample of 11 cartons of sugar for which

$$m = 1.040 \text{ kg}, \quad s = 0.0311 \text{ kg}$$

and the number of degrees of freedom available for the calculation of s is $\nu = 10$. This is an example of a small sample for which it is necessary to use the percentage points of the t-distribution given in table A5 instead of the normal distribution. For $\nu = 10$, the $P = 5$ % value of the t-distribution is found to be 2.23. This means that the 95 % confidence limits are

$$m \pm 2.23s/\sqrt{n} = 1.040 \pm 2.23 \times 0.0311/\sqrt{n}$$
$$= 1.040 \pm 0.021.$$

Thus there is a probability of $\frac{1}{20}$ of the mean of the whole output of the automatic packing machine being outside the range 1.019 to 1.061 kg.

It will be noticed that, as ν increases indefinitely, the $P = 5\%$ and $P = 0.2\%$ values of the t-distribution approach the values 1.96 and 3.09 used in § 106 for large samples.

108. Exercises.

1. (*a*) Obtain the 99.8 % confidence limits for machine A, §6, Ex. 2.

(*b*) What is the probability of the mean of the whole output being between 1.025 and 1.055?

2. (*a*) Obtain the 99.9 % confidence limits of the mean thickness of the lead cover of cable A, §6, Ex. 3.

(*b*) What is the probability of another sample of 10 measurements having a mean (i) outside the limits 0.745 ± 0.012, (ii) greater than 0.757?

3. From a large consignment of glass bottles a random sample of 400 bottles is drawn, their volumes being measured. The mean and standard deviation of these volumes are 507.3 c.c. and 8.2 c.c. respectively. Estimate the standard error of the mean and hence derive limits which have a 49 to 1 chance of including the mean volume of the whole consignment. How big would a sample have to be to make such limits differ by 1 c.c., the sample being drawn from a consignment with the same standard deviation as before? [Northern]

109. The significance of a single mean (large samples). In § 106 we saw that, in 95 % of cases,

$$m - 1.96\sigma/\sqrt{n} < \text{true } \mu < m + 1.96\sigma/\sqrt{n}.$$

If a given value of the mean, μ_0, lies in the above range we can accept it as a possible value of the true μ with 95 % confidence; if it does not lie in the range we can reject it with 95 % confidence. This statement holds provided

(i) the distribution of μ is normal, and

(ii) σ is known.

It is approximately true if n is large and s is used for σ. Thus we ask if

$$m - 1.96s/\sqrt{n} < \mu_0 < m + 1.96s/\sqrt{n}$$

and this is the same as asking if

$$\frac{|\mu_0 - m|}{s/\sqrt{n}} > 1.96.$$

Thus *the mean m of a large sample is said to differ significantly from a given value μ_0 if*

$$\frac{|\mu_0 - m|}{s/\sqrt{n}} > 1.96.$$

The difference between μ_0 and m in this case is said to be *at the 5 % level of significance.*

If

$$\frac{|\mu_0 - m|}{s/\sqrt{n}} > 3.09$$

the difference between μ_0 and m is at the 0.2 % level of significance; that is to say *very significant indeed* because the probability of the mean of a sample differing from the true mean by an amount large enough to make this possible is $\frac{1}{500}$.

110. Example. *The average breaking strength of steel rods is specified as* 20 *tons. The breaking strengths of* 100 *rods when measured are found to have a mean of* 19.9 *tons with a standard deviation of* 0.4 *tons. Is a complaint that the rods are not up to specification statistically justified?*

Taking $\mu_0 = 20$, $m = 19.9$, $s = 0.4$ and $n = 100$,

$$\frac{|\mu_0 - m|}{s/\sqrt{n}} = \frac{0.1\sqrt{100}}{0.4}$$

$$= 2.5.$$

Thus the mean of the sample differs significantly from the specified average and the complaint that the rods are not up to specification is justified.

111. Exercises.

1. A machine producing components to a nominal dimension of 2.000 cm is reset every morning. The first 50 components produced one morning have a mean of 2.001 cm with standard deviation 0.003 cm. Does this provide sufficient evidence that the machine is set too high?

2. Suppose that in the example of §110 only 50 rods had been tested and that the mean and standard deviation were the same. Would the complaint be justified?

3. Suppose that, in the example of §110, 100 rods were tested and that although the mean was the same 19.9 tons the standard deviation was 0.6 tons. Would the complaint be justified?

112. The significance of a single mean (small samples). The significance of the mean, m, of a small sample is tested by comparing

$$t = \frac{|\mu_0 - m|}{s/\sqrt{n}}$$

with the percentage points of the t-distribution given in table A5. As in §89, μ_0 is the mean of the whole population of which the sample is assumed to be a part. It will be noticed that the percentage points of the t-distribution depend on the number of degrees of freedom, ν, which are available for the calculation of the estimated standard deviation of the whole population and by §7

$$\nu = n - 1.$$

The following example illustrates the method of applying the t-test. *Large samples, equal in size, of male and female plants of dog's mercury were collected at each of* 13 *sites in Derbyshire. The mean numbers of leaf pairs in these samples were as follows:*

Site	1	2	3	4	5	6	7	8	9	10	11	12	13
Male	6.0	6.6	8.0	7.0	6.4	6.9	6.1	6.9	6.6	8.2	7.9	7.0	7.5
Female	5.8	7.3	6.6	6.7	6.3	6.2	6.1	7.3	6.0	6.7	8.2	6.0	6.5

Find the differences between the mean numbers of leaf pairs of male and female plants from site to site, and calculate the mean and the standard deviation of these differences.

State whether or not these data can be taken to establish a real difference between the average numbers of leaf pairs for male and female plants.

[Northern]

The required differences are:

0.2, −0.7, 1.4, 0.3, 0.1, 0.7, 0.0, −0.4, 0.6, 1.5, −0.3, 1.0, 1.0.

The mean of these differences is

$$m = 5.4/13$$
$$= 0.4154$$

and the standard deviation, which is an estimate of the standard deviation, σ, of the whole population of differences,

$$s = \surd(5.697/12)$$
$$= 0.6891.$$

If we assume that there is no real difference between the average numbers of leaf pairs for male and female plants we should expect the mean of the differences to be zero. We therefore take

$$\mu_0 = 0, \quad m = 0.4154, \quad s = 0.6891, \quad n = 13$$

and obtain

$$t = \frac{|\mu_0 - m|}{s/\surd n}$$
$$= 2.173.$$

By table A 5, when $\nu = 12$, the $P = 5\%$ value of t is 2.18. As the calculated value of t, 2.173, is not greater than the $P = 5\%$ value, the mean of the differences does not differ significantly from zero. This means that the data cannot be taken to establish a real difference between the average numbers of leaf pairs for male and female plants.

Note that the value of t is so near to the $P = 5\%$ value that further investigation seems desirable. If collections were to be made at a greater number of sites, say 21 altogether, and the mean and standard deviation were found to be unaltered, a real difference would then be established because the value of t would be $0.4154\sqrt{21}/0.6891$ and the $P = 5\%$ value of the t-distribution when $\nu = 20$ is 2.09.

The example just discussed illustrates the *method of controls*. It may be possible to test the significance of an effect by *choosing the individuals in pairs*. Thus two groups are formed each consisting of an individual from each pair. Then

either (i) one of the groups is subjected to a treatment under test, the other group (the *control*) being left untreated;

or (ii) the two groups are subjected to two different treatments which are being compared.

A third alternative is that each individual provides a pair of readings. In any of these cases the *mean of the differences between the pairs of observations* is tested for significance against a hypothetical mean of zero. Further examples of this method are: §126, Exs. 6, 12, 14, 15, 16, 18 and chapter 15, Ex. 47.

113. Exercises.

1. The average breaking strength of steel rods is specified as 20 thousand lb. A random sample of 10 rods had the following breaking strengths (in thousands of lb.):

23 21 16 24 19 22 25 19 18 24.

Investigate the significance of the mean.

2. A machine producing components to a nominal dimension of 1.125 is reset each morning. The first half-hour's production one morning is:

1.125 1.127 1.131 1.124 1.128 1.126 1.126 1.127 1.132 1.128 1.124 1.126.

Does this provide convincing evidence that the machine is set high?

114. The variance of the sums = the sum of the variances. The variance of the differences = the SUM of the variances.

Suppose we have 4 red cards numbered 6, 7, 9, 10 and 3 black cards numbered 1, 3, 5. The red and black cards represent two independent populations. If we select at random 1 red card and 1 black the 12 possible pairs of numbers we might obtain are:

6 and 1, 6 and 3, 6 and 5, 7 and 1, 7 and 3, 7 and 5, 9 and 1,
9 and 3, 9 and 5, 10 and 1, 10 and 3, 10 and 5.

Now the mean and variance of the 4 *red numbers* are 8 and 10/4 respectively and the mean and variance of the 3 *black numbers* are 3 and 8/3 respectively. Note that in this case we are not estimating the variance of a population from that of a sample and therefore we use $\Sigma\{(x-\bar{x})^2/n\}$ and not $\Sigma\{(x-\bar{x})^2/(n-1)\}$. Further, the mean of the sums of the pairs of numbers is

$$\tfrac{1}{12}\{7+9+11+8+10+12+10+12+14+11+13+15\} = 11$$

and the variance is

$$\tfrac{1}{12}\{16+4+0+9+1+1+1+1+9+0+4+16\} = \tfrac{62}{12}.$$

Thus *the mean of the sums = the sum of the means*

and *the variance of the sums = the sum of the variances.*

Also, the mean of the differences of the pairs of numbers is

$$\tfrac{1}{12}\{5+3+1+6+4+2+8+6+4+9+7+5\} = 5$$

and the variance is

$$\tfrac{1}{12}\{0+4+16+1+1+9+9+1+1+16+4+0\} = \tfrac{62}{12}.$$

Thus *the mean of the differences = the difference of the means*

and *the variance of the differences = the* SUM *of the variances.*

The above special cases are examples of two important general theorems which may be stated as follows:

If the mean and variance of the m values $x_1, x_2, x_3, \ldots, x_m$ *are* $\bar{x}$ *and* s_x^2 *and the mean and variance of the n values* $y_1, y_2, y_3, \ldots, y_n$ *are* $\bar{y}$ *and* s_y^2 *then* (i) *the mean and variance of the mn values of z given by* $z = x+y$ *are* $\bar{x}+\bar{y}$ *and* $s_x^2+s_y^2$, *and* (ii) *the mean and variance of the mn values of z given by* $z = x-y$ *are* $\bar{x}-\bar{y}$ *and* $s_x^2+s_y^2$. (It being understood that x and y represent two independent populations.)

115. Experiment with two coloured dice. If a single die is thrown a large number of times the scores 1, 2, 3, 4, 5, 6 are all equally probable and the mean score will therefore approximate to

$$\tfrac{1}{6}\{1+2+3+4+5+6\} = 3\tfrac{1}{2}. \qquad (1)$$

While the variance of the scores will approximate to

$$\tfrac{1}{6}(6\tfrac{1}{4}+2\tfrac{1}{4}+\tfrac{1}{4}+\tfrac{1}{4}+2\tfrac{1}{4}+6\tfrac{1}{4}) = 2\tfrac{11}{12}. \qquad (2)$$

If a white die and a black die are thrown together a large number of times and a record is kept of

(*a*) the white score + the black score for each throw,
(*b*) the white score − the black score for each throw,

§114 leads us to expect:

the mean of the sums (*a*) = 7, (3)
the variance of the sums (*a*) = $5\tfrac{5}{6}$, (4)
the mean of the differences (*b*) = 0, (5)
the variance of the differences (*b*) = $5\tfrac{5}{6}$. (6)

The facts of (1), (2), (3), (4), (5) and (6) above should be tested experimentally by throwing a single die

(i) 10 times, (ii) 50 times, (iii) 100 times,

and by throwing a pair of differently coloured dice

(i) 10 times, (ii) 50 times, (iii) 100 times.

116. Three or more dice. If the ideas of §§114 and 115 are extended to the mean and variance of the total score when three or more dice are thrown together the results shown in table 9B hold:

TABLE 9B

No. of dice thrown together	Mean score	Variance of scores	Standard deviation of scores
1	$3\frac{1}{2}$	$2\frac{11}{12}$	1.708
2	7	$5\frac{5}{6}$	2.416
3	$10\frac{1}{2}$	$8\frac{3}{4}$	2.958
4	14	$11\frac{2}{3}$	3.416
5	$17\frac{1}{2}$	$14\frac{7}{12}$	3.819

An important generalisation of the above is:

If $x_1, x_2, x_3, \ldots, x_n$ are n statistically independent variables

(i) *the mean of* $(x_1+x_2+x_3+\ldots+x_n)$ = *mean of* x_1 + *mean of* x_2 + *mean of* $x_3+\ldots+$ *mean of* x_n,

(ii) *the variance of* $(x_1+x_2+x_3+\ldots+x_n)$ = *variance of* x_1 + *variance of* x_2 + *variance of* $x_3+\ldots+$ *variance of* x_n.

A practical application follows.

117. Example. *In a certain workshop electrical resistances of three types are manufactured. Type A has a mean resistance of* 50 *ohms and standard deviation* 2 *ohms, type B a mean resistance of* 10 *ohms and standard deviation* 0.5 *ohms, type C a mean resistance of* 5 *ohms and standard deviation* 0.3 *ohms, each type being normally distributed about its mean. Resistances of approximately* 85 *ohms are then constructed by connecting in series one of type A, three of type B and one of type C. The resistances so constructed can be accepted only if accurate measurement shows them to be between* 80 *and* 90 *ohms; otherwise they must be rejected. Calculate the percentage likely to be rejected. Show further that the probability that any one of the resistances is between* 84 *and* 86 *ohms is slightly more than* $\frac{1}{3}$. [Northern]

The mean of the *sum* of the resistances $= 50+10+10+10+5$

$= 85$

and the variance of the *sum* $= 2^2+(0.5)^2+(0.5)^2+(0.5)^2+(0.3)^2$

$$= 4.84.$$

Moreover, it can be proved that if the separate resistances are normally distributed, the *sum* is normally distributed. Thus the standard deviation of the sum = 2.2 and the standardised deviate of 80 or 90 from the mean is $5/2.2 = 2.273$. Now, by table A2, $A(2.273) = 0.98849$ and hence the percentage of 85 ohm resistances below 80 ohms or above 90 ohms is $100-98.849 = 1.151$. Thus the percentage likely to be rejected is 2.30.

Further, the standardised deviate of 84 or 86 from the mean is

$$1/2.2 = 0.4545 \quad \text{and since} \quad A(0.4545) = 0.6752$$

it follows that 17.52 % of the resistances are between 85 and 86 and 17.52 % between 84 and 85. Thus 35.04 % are between 84 and 86 and this is equivalent to the statement that the probability of any one of the resistances being between 84 and 86 ohms is slightly more than $\frac{1}{3}$.

118. Exercises.

1. A certain firm mass-produced machines which, in the course of their assembly passed through four workshops, *A*, *B*, *C* and *D*. A record of the times taken in each workshop and the times of transit from one workshop to the next was kept and from it the following summary showing the means and standard deviations of the times was published:

	Mean time (h)	Standard deviation (h)
Workshop *A*	3.48	0.25
Transit from *A* to *B*	0.23	0.05
Workshop *B*	4.56	0.30
Transit from *B* to *C*	0.53	0.12
Workshop *C*	1.91	0.20
Transit from *C* to *D*	0.32	0.10
Workshop *D*	2.67	0.20

Assuming that the workshop and transit times were independently normally distributed, calculate the mean and standard deviation of the times taken for the complete assembly of the machines, and show that only 1 % of the machines were assembled in less than $12\frac{1}{2}$ h while 6 % took over $14\frac{1}{2}$ h. [Northern]

2. Four athletes specialise in running 200 m, 200 m, 400 m and 800 m respectively. They train as a team for a 1600 m medley relay race. During training their mean times for their respective distances are 23.9 sec, 24.1 sec, 53.6 sec and 2 min 7.4 sec, and the corresponding standard deviations are 0.3 sec, 0.3 sec, 0.8 sec and 1.8 sec. Assuming that their individual times are normally and independently distributed, estimate the mean and standard deviation of the times in which the team covers 1600 m. Also estimate the probability that, on any particular occasion, the time will be 3 min 45 sec or less. [Northern]

3. At the assembly stage in the manufacture of a knitting machine, four sections of bakelite of lengths x_1, x_2, x_3, x_4 are drawn at random from a box containing a large number of these sections and fitted between two pieces of metal on the partially assembled machine, these being separated by a distance y as shown in the diagram. The fit is considered satisfactory if the clearance $y-x_1-x_2-x_3-x_4$ lies between zero and 0.10 cm (see fig. 12).

If the lengths of the pieces in the box vary randomly about a mean of 4.00 cm, with a standard deviation of 0.015 cm, and if y varies randomly from machine to machine with mean 16.06 cm and standard deviation 0.024 cm, find the mean and standard deviation of the distribution of $y-x_1-x_2-x_3-x_4$.

Hence determine the proportion of assemblies giving a satisfactory fit at the first attempt. [Northern]

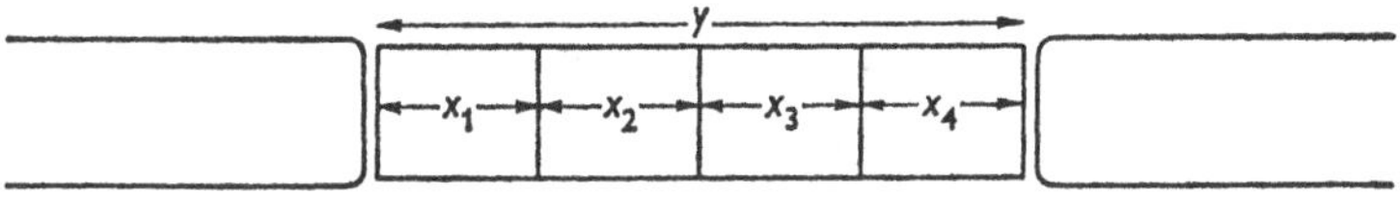

Fig. 12

119. Var$(a_1x_1+a_2x_2+\ldots+a_nx_n) = a_1^2\text{var}\ x_1+a_2^2\text{var}x_2+\ldots+a_n^2\text{var}x_n$. The following example extends still further the general principles of §116.

Three dice, each numbered in the usual way from one to six, are coloured white, red and blue respectively. After casting them a boy 'scores' in the following way. To the white number he adds twice the red number and then subtracts the blue number. Thus a white three, a red four and a blue two would score

$$3+8-2 = 9.$$

Assuming that the boy casts the dice a large number of times calculate the mean and variance of the scores. [Northern]

Representing the white number by x_1, the red by x_2 and the blue by x_3, the method of scoring is

$$x_1+2x_2-x_3.$$

Now it can be proved mathematically that if x_1, x_2, x_3 are statistically independent variables and a_1, a_2, a_3 are constants then (i) the mean of $(a_1x_1+a_2x_2+a_3x_3) = a_1(\text{mean of } x_1)+a_2(\text{mean of } x_2)+a_3(\text{mean of } x_3)$, and (ii) the variance of $(a_1x_1+a_2x_2+a_3x_3) = a_1^2(\text{var } x_1)+a_2^2(\text{var } x_2)+a_3^2(\text{var } x_3)$. In the example under consideration

$$a_1 = 1, \quad a_2 = 2, \quad a_3 = -1,$$

the mean of x_1 = the mean of x_2 = the mean of x_3 = $3\frac{1}{2}$,
the variance of x_1 = the variance of x_2 = the variance of x_3 = $2\frac{11}{12}$.
Thus the mean score $= 1\times 3\frac{1}{2}+2\times 3\frac{1}{2}+(-1)\times 3\frac{1}{2}$
$= 7$

and the variance of the scores $= 1^2\times 2\frac{11}{12}+2^2\times 2\frac{11}{12}+(-1)^2\times 2\frac{11}{12}$
$= 17\frac{1}{2}$.

120. Example. *A large consignment of mercury is supplied in small containers. The volumes of mercury in the containers are distributed about a mean of* 250 *c.c. with a standard deviation of* 2 *c.c. and the weights of the empty containers are distributed about a mean of* 456 g *with a standard deviation of* 7 g. *Calculate the mean and standard deviation of the full containers given that the specific gravity of mercury is* 13.6.

Taking the mean of $x_1 = 250$, the mean of $x_2 = 456$,

$$\text{the variance of } x_1 = 4, \text{ the variance of } x_2 = 49,$$

$$a_1 = 13.6, \quad a_2 = 1$$

we obtain (i) the mean weight of the full containers is

$$13.6 \times 250 + 1 \times 456 = 3856 \text{ g}$$

and (ii) the variance of these weights is

$$13.6^2 \times 4 + 1^2 \times 49 = 789 \text{ g}^2.$$

Thus the standard deviation is 28.1 g.

121. The significance of the difference between the means of two large samples. Consider the two large samples details of which are given in table 9C.

TABLE 9C

	Sample 1	Sample 2
No. of observations	n_1	n_2
Mean	m_1	m_2
Standard deviation	s_1	s_2
Standard error	$s_1/\sqrt{n_1}$	$s_2/\sqrt{n_2}$

If we were to write down all the possible differences between the observations of sample 1 and sample 2 in the same way that we wrote down the differences of the red and black numbers in §114, the variance of the population of differences is the sum of the variances of the separate populations. Now the estimates of the variances of the separate populations are s_1^2/n_1 and s_2^2/n_2 and hence the estimated variance of the population of differences is $\frac{s_1^2}{n_1}+\frac{s_2^2}{n_2}$. Thus, *the standard error of the difference is*

$$\sqrt{\left(\frac{s_1^2}{n_1}+\frac{s_2^2}{n_2}\right)}.$$

Moreover, the mean of the differences being equal to the difference of the means is (m_1-m_2) and if sample 1 and sample 2 both belong to the same population, the mean of the differences should not differ significantly from zero. Hence, for large samples, if

$$\frac{|m_1 - m_2|}{\sqrt{\left(\frac{s_1^2}{n_1} + \frac{s_2^2}{n_2}\right)}} > 1.96,$$

the difference between the means is significant at the 5 % level. This means that the chance of such a large 'difference' when the samples are from the same parent population is less than $\frac{1}{20}$.

If

$$\frac{|m_1 - m_2|}{\sqrt{\left(\frac{s_1^2}{n_1} + \frac{s_2^2}{n_2}\right)}} > 3.09,$$

the difference between the means is significant at the 0.2 % level and the chance of such a large 'difference' when the samples are from the same parent population is $\frac{1}{500}$.

122. Example. *Sixty boys who entered a school A and sixty boys who entered another school B were given the same examination in English. After each group of boys had attended their respective schools for one year they were each given another common examination in English. The means and standard deviations of the marks are shown in the following table:*

	Examination mark of the 60 boys in each group			
	Upon entry		After one year in the school	
	Mean	Standard deviation	Mean	Standard deviation
School *A*	53	10	51	7
School *B*	50	10	48	7

Show that the difference between the means was not significant at the 5 % level when the boys entered the schools but that it was significant at the 5 % level after one year. [Northern]

Interpret this result.

When the boys entered the two schools,

$$\frac{|m_1 - m_2|}{\sqrt{\left(\frac{s_1^2}{n_1} + \frac{s_2^2}{n_2}\right)}} = \frac{(53-50)}{\sqrt{\left(\frac{10^2}{60} + \frac{10^2}{60}\right)}}$$

$$= 1.64.$$

Thus the difference between the means was not significant at the 5 % level.

After one year,

$$\frac{|m_1-m_2|}{\sqrt{\left(\frac{s_1^2}{n_1}+\frac{s_2^2}{n_2}\right)}} = \frac{(51-48)}{\sqrt{\left(\frac{7^2}{60}+\frac{7^2}{60}\right)}}$$

$$= 2.35$$

and the difference between the means was then significant at the 5 % level. This means that when the two groups of boys entered the schools they both belonged to the same parent population. Although the mean for school *A* was 3 higher than that for school *B* the variability of the individuals was so great that the difference of 3 could be regarded as nothing more than a chance effect. The members of school *A* could not be considered more able to cope with the examination than those of school *B*. After one year, however, the two groups no longer belonged to the same parent population. The variability within the groups was reduced to such an extent that the difference of 3 could no longer be regarded as a chance effect. The tuition given in school *A* had been a better preparation for the examination than that given in school *B*.

By table A2, $A(2.35) = 0.99061$ which is 99 % approximately. This shows that the value, 2.23, of the S.E. is such that the two-tail probability (§104) of the samples being from the same parent population is 2×1 %. Thus, after one year, the difference between the means is significant at the 2 % level.

123. Exercises.

1. In order to find out whether the average speed of motor-vehicles leaving London is different from that of motor-vehicles entering London, cars and motor-cycles were timed over a stretch of the Portsmouth road. The following table shows the results of the investigation:

	Leaving London	Entering London
No. of vehicles timed	50	50
Mean time in sec	17.04	18.38
Variance in sec.^{2}	8.846	9.106

Determine whether the difference between the means is significant (i) at the 5 % level, (ii) at the 1 % level. Comment on the meaning of your result.

[Northern]

2. A firm which manufactures lead-covered submarine cable suspected that the lead was being put on more thickly by its night workers than by its day workers. To keep down the cost the lead must be as thin as possible but must be nowhere less than 0.2 in. thick if it is to withstand the action of the sea water and general wear and tear. Part of an investigation carried out is summarised in the following table:

	Day work	Night work
No. of places at which the thickness of the lead cover was measured	100	100
Mean thickness (in.)	0.292	0.298
Standard deviation (in.)	0.021	0.019

Determine whether the difference between the means is significant or not and make comments on the implications of the given figures and your result.

[Northern]

124. The significance of the difference between the means of two small samples. For small samples it is necessary to replace the expression

$$\frac{|m_1 - m_2|}{\sqrt{\left(\frac{s_1^2}{n_1} + \frac{s_2^2}{n_2}\right)}} \quad \text{by} \quad \frac{|m_1 - m_2|}{s\sqrt{\left(\frac{1}{n_1} + \frac{1}{n_2}\right)}},$$

where s is an estimate of the standard deviation of the parent population from which both samples are drawn. The procedure for an investigation of the difference between the means of two small samples is as follows:

(i) Assume that the samples are from populations with the same variance, σ^2, even if the means differ. Estimate σ^2 by

$$s^2 = \frac{\Sigma(x_1 - m_1)^2 + \Sigma(x_2 - m_2)^2}{(n_1 - 1) + (n_2 - 1)} \quad \text{(if the individual observations are available)}$$

or $$s^2 = \frac{(n_1 - 1)s_1^2 + (n_2 - 1)s_2^2}{(n_1 + n_2 - 2)} \quad \text{(if the standard deviations are available but not the individual observations).}$$

Note that the number of degrees of freedom available for the calculation of s is $n_1 + n_2 - 2$.

(ii) Calculate the value of

$$t = \frac{|m_1 - m_2|}{s\sqrt{\left(\frac{1}{n_1} + \frac{1}{n_2}\right)}}.$$

If t is greater than the $P = 5\%$ value of the t-distribution given in table A5 for $\nu = n_1 + n_2 - 2$, the difference between the means is significant at the 5% level and the assumption or null hypothesis that the samples are from populations with the same mean is rejected.

Note that assumption (i) is usually valid because, if small samples are used, it is generally in relation to a carefully designed experiment, i.e. in §125, the variability is likely to be due to differences between individual rabbits, or slight inaccuracies in the administration of the drug, and so will be the same whether the standard or trial preparation is used.

125. Example. *In the manufacture of insulin, the strength of the final product may be checked by making a comparison between the mean level of blood sugar in a group of rabbits inoculated with it and the mean level of blood sugar in a comparable group of rabbits inoculated with a standard preparation of known potency. The results of such a test were as follows:*

Standard preparation: 36, 61, 60, 63, 57, 58, 61, 48, 54, 75, 68, 65.

Trial preparation: 58, 58, 76, 63, 50, 54, 63, 64, 65, 87, 72, 80.

Calculate the difference between the two means and estimate its standard error. Verify that the difference is not statistically significant.

Can it be concluded that the trial and standard preparations have equal potency? Give reasons for your answer. [Northern]

Table 9D shows the calculation of

$$\Sigma x_1 = 706, \quad \Sigma x_2 = 770, \quad \Sigma x_1^2 = 42614 \quad \text{and} \quad \Sigma x_2^2 = 50532.$$

TABLE 9D

	Standard preparation x_1	Trial preparation x_2	x_1^2	x_2^2
	36	58	1296	3364
	61	58	3721	3364
	60	76	3600	5776
	63	63	3969	3969
	57	50	3249	2500
	58	54	3364	2916
	61	63	3721	3969
	48	64	2304	4096
	54	65	2916	4225
	75	87	5625	7569
	68	72	4624	5184
	65	60	4225	3600
Total	706	770	42614	50532

Thus

$$m_1 = \frac{\Sigma x_1}{n_1} = 706/12 = 58.83, \qquad m_2 = \frac{\Sigma x_2}{n_2} = 770/12 = 64.17,$$

$$(n_1-1)s_1^2 = \Sigma(x_1-m_1)^2 = \Sigma x_1^2-(\Sigma x_1)^2/n_1 = 42614-706^2/12 = 1078,$$

$$(n_2-1)s_2^2 = \Sigma(x_2-m_2)^2 = \Sigma x_2^2-(\Sigma x_2)^2/n_2 = 50532-770^2/12 = 1124.$$

Hence $$s^2 = \frac{(n_1-1)s_1^2+(n_2-1)s_2^2}{(n_1+n_2-2)}$$

$$= \frac{1078+1124}{22}$$

$$= 100.1$$

and $$s = 10.0.$$

The estimate of the S.E. of the difference between the means is, therefore,

$$s\sqrt{\left(\frac{1}{n_1}+\frac{1}{n_2}\right)} = 10.0\sqrt{\left(\frac{1}{12}+\frac{1}{12}\right)}$$

$$= 4.08$$

and $$t = \frac{|m_1-m_2|}{s\sqrt{\left(\frac{1}{n_1}+\frac{1}{n_2}\right)}}$$

$$= (64.17-58.83)/4.08$$

$$= 1.31.$$

Now by table A5, the $P = 5\%$ value of the t-distribution for $\nu = 22$ is between 2.06 and 2.09. As the calculated value of t is not greater than the $P = 5\%$ value the difference between the means is not significant and the null hypothesis is not rejected.

The null hypothesis is that both samples are drawn from the same parent population. That is to say, the trial preparation and the standard preparation are of equal potency.

126. Exercises.

1. For a random sample of 16 households from one district the sum (in shillings) of weekly rents is 190 and the sum of the squares 3059. The corresponding figures for 26 households from a second district are 340 and 6108. Estimate the mean and variance of rents in each of the two districts.

Estimate also the standard error of the difference between the two means, and test whether or not the difference is statistically significant. [Northern]

2. Ten soldiers visit the rifle range two weeks running. The first week their scores are

67, 24, 57, 55, 63, 54, 56, 68, 33, 43.

The second week they score, in the same order,

70, 38, 58, 58, 56, 67, 68, 77, 42, 38.

Is there any significant improvement? How would the test be affected if the scores in the second week were for a different group of soldiers? [A.I.S.]

3. A standard cell, whose voltage is known to be 1.10 volts, was used to test the accuracy of two voltmeters, *A* and *B*. Ten independent readings of the voltage of the cell were taken with each voltmeter, and the results were as follows:

A	1.11	1.15	1.14	1.10	1.09	1.11	1.12	1.15	1.13	1.14
B	1.12	1.06	1.02	1.08	1.11	1.05	1.06	1.03	1.05	1.08

From these results is there any evidence of bias in either voltmeter? [R.S.S.]

4. A group of 7 seven-week-old chickens, reared on a high-protein diet, weigh 12, 15, 11, 16, 14, 14, 16 ounces; a second group of 5 chickens, similarly treated except that they receive a low-protein diet, weight 8, 10, 14, 10, 13 ounces.

Calculate the value of t and test whether there is significant evidence that additional protein has increased the weight of the chickens. Criticise the arrangement of the experiment and suggest improvements in design. [R.S.S.]

5. A group of 8 psychology students were tested for their ability to remember certain material, and their scores (number of items remembered) were as follows:

A	*B*	*C*	*D*	*E*	*F*	*G*	*H*
19	14	13	16	19	18	16	17

They were then given special training purporting to improve memory and were re-tested after a month.

Their scores were then:

A	*B*	*C*	*D*	*E*	*F*	*G*	*H*
26	20	17	21	23	24	21	18

A control group of 7 students was also tested and re-tested after a month, but was not given special training. The scores in the two tests were:

J	*K*	*L*	*M*	*N*	*O*	*P*
21	19	16	22	18	20	19
21	23	16	24	17	17	16

Compare the change in each of the two groups by calculating t and test whether there is significant evidence to show the value of the special training.

Is there evidence that the experiment was not properly designed?

[R.S.S.]

6. Sickness rates for two factories over a period of 6 months were as follows:

	Rate per 100	
	Factory *A*	Factory *B*
January	64	75
February	72	83
March	79	74
April	58	67
May	49	52
June	40	46

By means of the t-test, or otherwise, examine whether there is any significant difference between the two factories. [A.I.S.]

7. From each of two batches of electric lamps made by the same manufacturer, a random sample of six was chosen and tested by burning out. The results were as follows:

Length of life (h)	Batch *A*	802	959	1022	1040	733	897
	Batch *B*	839	961	1035	896	994	950

Is there any significant difference in average length of life between the two batches?

The average length of life should be at least 1000 h to satisfy a certain specification. Is there any reason to suppose, from the given data, that the manufacturer's product is not likely to meet this specification? [R.S.S.]

8. Metal rods are delivered in large batches to a customer. When production is under control, the rods have a mean diameter of 2.132 cm, with a standard deviation of 0.001 cm. A random sample of 90 from a batch is found to have a mean diameter of 2.1316 cm. Report on the likelihood that the production is still under control. [A.E.B.]

9. The standard deviation of the number of articles produced in an hour by a factory worker is 14 for all workers, but the mean number produced varies from worker to worker. The numbers produced by a new employee in a representative hour on each of ten successive days were: 108, 124, 92, 113, 129, 146, 117, 103, 131, 128. Do these values show that this worker's rate of production is, at the 5 % level of significance, below the factory average of 127 articles per hour?

Determine the two factory average rates from which the average of these ten observations would be judged just to differ at the 5 % level of significance. How many more observations would be needed to reduce to 14 the difference between the two rates defined in this way? [Cambridge]

10. A tractor drawbar is intended to have a breaking strain of 12 tons. It is known that, because of small variations in the casting, the breaking strain varies between individual bars with standard deviation 0.4 tons. A sample of 5 bars gave a mean breaking strain of 11.5 tons; is there significant evidence (at the 5 % level) of a change in breaking strain?

Obtain a limit which the new mean breaking strain may be said with 95 % confidence to exceed. [Cambridge]

11. In a test on electric light bulbs, 200 of the type I are found to have a mean life of 1500 hours with a standard deviation of 80 hours and 300 of type II are found to have a mean life of 1520 hours with a standard deviation of 90 hours. Is there any evidence that type II has a longer mean life than type I? Which type is more likely to fail in the first 1300 hours? State any assumptions you make about the distributions. [A.E.B.]

12. The times taken by ten men to perform a task were measured. After training, designed to reduce the time, the men were tested again. The times in seconds were found to be as follows:

Man no.	1	2	3	4	5	6	7	8	9	10
Time before	27	45	51	33	49	22	23	35	41	25
Time after	22	37	45	26	44	20	16	35	42	23

Report on the effectiveness of the training as shown by this experiment. [A.E.B.]

13. The barometric pressures in millibars at two observatories A and B, taken at random times, are as follows:

A	1002	996	998	1027	1031	1015	991
B	972	984	987	999	979	965	

Examine the hypotheses that the mean barometric pressure is the same at A as at B.

Suggest a better way of making observations to test this hypothesis. [A.E.B.]

14. Eight typists were given electric typewriters instead of their previous machines and the increases in the numbers of finished quarto pages in a day were found to be as follows:

$$8, \ 12, \ -3, \ 4, \ 5, \ 13, \ 11, \ 0.$$

Do these figures indicate that the typists in general can work faster with electric typewriters? State the assumptions you make. [A.E.B.]

15. Ten wooden panels are painted partly with an ordinary paint and partly with a flame-resisting paint. The times, in suitable units, taken to burn through the panels are as follows:

Panel	A	B	C	D	E	F	G	H	I	J
Ordinary	12	10	8	20	6	13	7	9	25	15
Flame resisting	17	16	15	22	13	20	13	18	24	18

Examine the hypothesis that the paint does not make any real difference, stating any assumptions you make. [A.E.B.]

16. Ten children were given an intelligence test I and, after some practice in similar tests, were given a test II of the same standard as I. Their scores in the two tests were as follows:

Child	A	B	C	D	E	F	G	H	I	J
Test I	40	32	20	63	77	51	43	38	37	82
Test II	45	39	27	72	85	54	45	37	39	88

Does the experiment provide evidence that the practice has made a difference? Give reasons for your answer. [A.E.B.]

17. In an experiment to compare two poultry rations 18 hens were kept in a battery for 8 weeks and the two rations were each fed to 9 birds. The numbers of eggs laid by the individual birds were

Ration 1	38	41	44	39	42	40	45	43	40
Ration 2	37	32	40	34	38	33	39	36	38

Carry out a t-test to compare the mean numbers of eggs on the two rations.

What difference between sample means is just significant at the 5 % level for samples of the same size and variance as those in the above table? [Cambridge]

18. In an experiment to compare two methods of rearing veal calves eight pairs of identical twins were used, one twin of each pair being allocated at random to each method of rearing. At the end of the experiment the calves were slaughtered and sample joints were cooked and scored for palatability with the following results:

Twin pair ...	1	2	3	4	5	6	7	8
Method *A*	27	37	31	38	29	35	41	37
Method *B*	23	28	30	32	27	29	36	31

Carry out a paired-sample t-test to decide whether the methods differ in mean palatability score; state the assumptions necessary for the validity of your test. [Cambridge]

19. The heights (in cm) of small random samples of boys and girls of age 14 were as follows:

Boy	160	158	159	162	164	157	156	
Girl	157	159	152	151	159	160	157	156

Carry out a 2-sample t-test to test whether the mean heights of the boys and girls differ at the 5 % level of significance.

If you were asked to test whether the mean height of the boys differed from 160 cm, would you use a standard error estimate based only on the boys' heights or, as in the test just performed, one based on the heights of all the children? Justify your answer. [Cambridge]

20. The breaking strain of a certain type of chain is expected to be 1.300 tons and the standard deviation is known to be 0.040 tons. The mean breaking strain of 9 test lengths was 1.323 tons; was there significant evidence of a *change* in breaking strain, and if so at what level of significance?

The manufacturing process was slightly modified, the standard deviation remaining unchanged: the first 9 test lengths gave breaking strains (tons) 1.254, 1.301, 1.344, 1.318, 1.394, 1.335, 1.262, 1.337, 1.362. Is there significant evidence that the breaking strain has been *increased*, and if so at what level of significance? [Cambridge]

21. An iron foundry asserts that the mean weight μ of the castings it produces is 20.0 kilos. The weights of a sample of 10 castings are:

19.8	20.3	20.6	21.1	19.3
19.6	20.1	20.8	21.1	21.3

Estimate μ and the standard deviation σ of the weight of a casting. Assuming that the true value of σ is equal to your estimated value and that the weight of a casting has a normal distribution, test whether the assertion concerning μ is contradicted significantly at the 5 % level of significance. [Cambridge]

22. A machine is making dimensioned parts and a random sample of ten are found to have the following lengths in cm:

1.504	1.496	1.492	1.501	1.503
1.505	1.495	1.500	1.493	1.501

Estimate the variance of the population
(*a*) from the sum of the squares, and
(*b*) by using the range.

By means of the t-distribution give 95 % confidence limits for the mean of the population. [A.E.B.]

10

QUALITY CONTROL

127. Control of a given dimension. When articles are being mass produced some variation in the dimensions must be expected and a certain tolerance limit is allowed in their specification. Thus, a dimension stated as 1.005 ± 0.005 in. has tolerance limits of 1.00 and 1.01 in. It is often possible, by random sampling, to ensure that the total output is within the tolerance limits. This application of statistics to industrial and manufacturing processes is known as *quality control*. Quality control is concerned with the collection, analysis and presentation of facts concerning quality. The quality control of a given dimension is a direct application of the normal distribution.

128. Control of the fraction defective. A second type of quality control is an application of the Poisson distribution. It is used when the articles being manufactured are not classified by dimensions but are deemed either *sound* or *defective* (*acceptable* or *unacceptable*) according as they either pass or fail a certain test of quality.

129. The value of random sampling. The method of judging the quality of the whole output by a system of random sampling is important not only because *it saves time and money*, but also because *the sample tested is often completely destroyed* in the test. The quality of the whole output of a munitions factory, for example, is tested by actually exploding random samples of its products.

TABLE 10A

Conversion of range to standard deviation (*Reproduced from* Lindley and Miller, *Cambridge Elementary Statistical Tables*, page 7)

n	a_n	n	a_n	n	a_n	n	n_n
2	0.8862	5	0.4299	8	0.3512	11	0.3152
3	0.5908	6	0.3946	9	0.3367	12	0.3069
4	0.4857	7	0.3698	10	0.3249	13	0.2998

An estimate of the standard deviation is given by multiplying the range of a random sample of size n, from a normal population, by a_n. The mean range in samples of size n from a normal population is the standard deviation of the population divided by a_n.

130. The estimation of the standard deviation from the mean range of samples of n observations. In industrial quality control, the standard deviation is not usually *calculated* from the sums of squares. Instead it is *estimated* by table 10A from the ranges of samples as shown in the following example.

131. Example. *Consider the* 5 *samples, each of* 4 *observations, given in* §103 *for which the calculated mean and standard deviation were found to be* 1.0055 *cm* and 0.00322 *cm respectively. Rewritten in thousandths of a cm above* 1.000, *the observations are as given in table* 10B:

TABLE 10B

Sample no. ...	1	2	3	4	5
	10	2	9	5	6
	2	7	7	11	2
	7	8	3	2	1
	7	5	0	8	8
Sample mean	$6\frac{1}{2}$	$5\frac{1}{2}$	$4\frac{3}{4}$	$6\frac{1}{2}$	$4\frac{1}{4}$
Range	8	6	9	9	7

The mean of the sample means is

$$\tfrac{1}{5}(6\tfrac{1}{2}+5\tfrac{1}{2}+4\tfrac{3}{4}+6\tfrac{1}{2}+4\tfrac{1}{4}) = 5.5 \text{ thousandths above } 1.000$$
$$= 1.0055 \text{ cm.}$$

The mean range is

$$\tfrac{1}{5}(8+6+9+9+7) = 7.8 \text{ thousandths.}$$

By table 10A, $a_4 = 0.4857$ and the estimated standard deviation is

$$7.8 \times 0.4857 = 3.787 \text{ thousandths}$$
$$= 0.0038 \text{ cm.}$$

132. A quality-control chart for means. Quality control of a given dimension is maintained by two charts, one showing the means of successive samples and the other their ranges (or standard deviations). In §102 we learned that, if the estimated standard deviation of the individual observations is σ, the estimated standard deviation of the means of samples of size n is $\sigma/\sqrt{n}$. Moreover, even though the individual observations are not normally distributed, it is likely that the means of samples are fairly normally distributed. Thus

(i) the probability of a sample having a mean outside the limits

$$\text{mean} \pm \frac{1.96\sigma}{\sqrt{n}} \quad \text{(the 95 \% zone)}$$

is 1 in 20,

(ii) the probability of a sample having a mean outside the limits

$$\text{mean} \pm \frac{3.09\sigma}{\sqrt{n}} \quad \text{(the 99.8\% zone)}$$

is 1 in 500.

In the example of the last paragraph

(i) the 95% zone is $1.0055 \pm 1.96 \times 0.0038/\sqrt{4} = 1.0055 \pm 0.0037$, i.e. between 1.002 and 1.009 (to the nearest thousandth);

(ii) the 99.8% zone is $1.0055 \pm 3.09 \times 0.038/\sqrt{4} = 1.0055 \pm 0.0059$, i.e. between 1.000 and 1.011 (to the nearest thousandth).

The quality-control chart for means in this case is shown in fig. 13. It is a graph with a pair of horizontal lines, $\bar{x} = 1.002$ and $\bar{x} = 1.009$, called the *inner control* lines and another pair, $\bar{x} = 1.000$ and $\bar{x} = 1.011$, called the *outer control* lines. If the mean of each random sample of 4 observations is

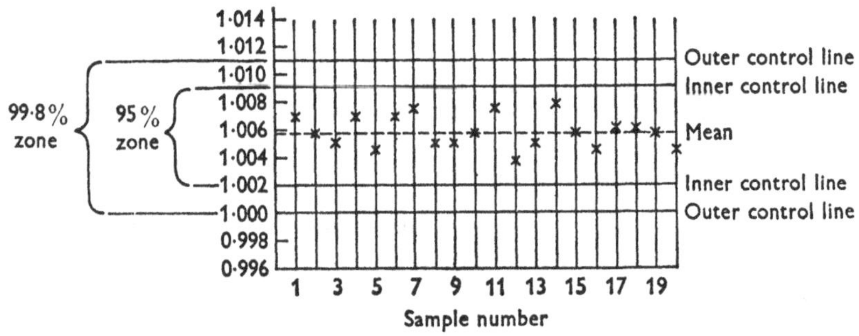

Fig. 13. A quality-control chart for means.

plotted on this chart and only 1 mean out of every 20 falls outside the inner control limits (and only 1 out of every 500 outside the outer control limits) *the production is said to be under control.* Immediately these proportions are exceeded *the production is out of control.*

In the example given, the estimates of the mean and the standard deviation were based on 5 samples only. This may be necessary if it is desired to get a system of control operating as quickly as possible, but it is advisable to make a second estimate of the mean and the standard deviation when another 10 or 20 samples are available and to readjust the control lines accordingly.

In fig. 13 the means of the 5 samples given in §128 have been plotted and also the means of the next 15 samples given in table 10C.

Figure 13 indicates that production was under control over the whole period during which the 20 samples were drawn.

TABLE 10C

Sample no. ...	6	7	8	9	10	11	12	13
Sample mean	$6\frac{1}{2}$	$6\frac{3}{4}$	5	5	$5\frac{1}{2}$	$6\frac{3}{4}$	$3\frac{1}{2}$	$5\frac{1}{4}$
Range	8	9	6	8	7	9	7	8

Sample no. ...	14	15	16	17	18	19	20
Sample mean	$7\frac{1}{2}$	$5\frac{1}{2}$	$4\frac{1}{4}$	$5\frac{3}{4}$	$5\frac{3}{4}$	$5\frac{1}{2}$	$4\frac{1}{2}$
Range	5	9	6	8	8	9	7

133. Exercise.

Mass-produced 100-*ohm electrical resistors. Random samples of* 5 *resistors drawn during the course of production* (*resistances given in ohms above* 90)

Sample no.	1	2	3	4	5	6	7	8	9	10
Beginning	9	8	14	13	9	11	5	7	9	10
of pro-	8	15	12	5	13	11	12	10	12	10
duction	6	15	14	12	8	9	12	8	10	12
	8	14	12	5	14	16	9	10	10	13
	13	11	13	11	9	14	15	9	15	13

Sample no.	11	12	13	14	15	16	17	18	19	20
After	17	6	9	17	11	14	6	12	10	11
1 week	17	13	9	7	13	14	9	7	4	12
	10	14	9	9	7	9	10	12	10	11
	8	13	16	10	7	13	14	15	11	7
	6	13	11	18	13	12	12	11	9	15

Sample no.	21	22	23	24	25	26	27	28	29	30
After	12	12	15	10	14	8	3	15	11	15
4 weeks	7	17	10	14	8	4	11	7	18	9
	14	13	13	17	3	16	11	1	15	10
	14	10	13	19	17	10	16	16	11	7
	9	10	15	17	16	15	8	15	11	12

Random samples nos. 1–10 above were drawn soon after production had begun, samples nos. 11–20 about 1 week later and samples nos. 21–30 about 4 weeks later. Use the first 10 samples to set up a quality-control chart for means by calculating

(i) the mean of the sample means,
(ii) the mean range,
(iii) the 95 % zone,
(iv) the 99.8 % zone.

Plot the means of samples nos. 11–20 on the chart and comment on the quality of production after 1 week.

Plot the means of samples nos. 21–30 on the chart and comment on the quality of production after 4 weeks.

134. A quality-control chart for ranges. By using table 10D a quality-control chart for ranges can be set up to ensure that variation within the samples is not too great. Thus for the 5 samples of §131 we take the estimate of σ, 3.787 thousandths, and obtain

(i) the 97.5 % zone for ranges,

$$D_{0.975} \times \sigma = 3.98 \times 3.787$$
$$= 15.1 \text{ thousandths,}$$

(ii) the 99.9 % zone for ranges,

$$D_{0.999} \times \sigma = 5.30 \times 3.787$$
$$= 20.1 \text{ thousandths.}$$

TABLE 10D

Distribution of ranges of samples

n	2	3	4	5	6	7	8	9	10	11	12
$D_{0.975}$	3.17	3.68	3.98	4.20	4.36	4.49	4.61	4.70	4.79	4.86	4.92
$D_{0.999}$	4.65	5.05	5.30	5.45	5.60	5.70	5.80	5.90	5.95	6.05	6.10

If σ is the standard deviation of the population of individual observations (i) 97.5 % of the ranges of samples of size n will not exceed $D_{0.975} \times \sigma$, (ii) 99.9 % of the ranges of samples of size n will not exceed $D_{0.999} \times \sigma$.

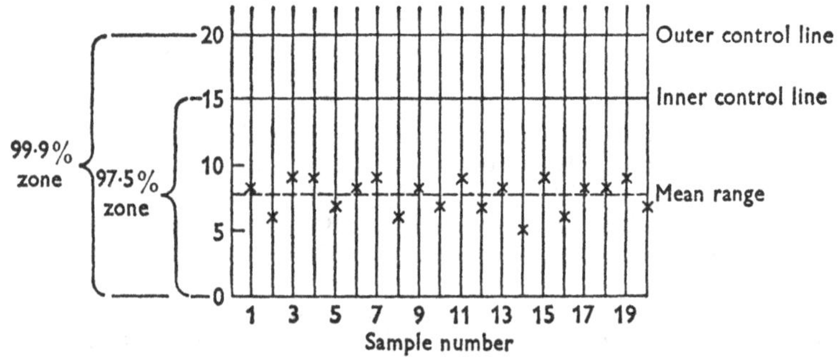

Fig. 14. A quality-control chart for ranges.

Figure 14 shows the quality-control chart. On it are plotted the ranges of samples nos. 1–5 given in §131 followed by samples nos. 6–20 given in §132. If only 1 range out of every 40 is above the inner control limit (or only 1 out of every 1000 above the outer control limit) the production is under control. If these proportions are exceeded the production is out of control. Figure 14 shows that for the example under consideration quality is being well maintained.

135. Exercise.

Using the first ten samples of §133, set up a quality-control chart for ranges.

Plot the ranges of samples nos. 11–20 on the chart and comment on production after 1 week.

Plot the ranges of samples nos. 21–30 on the chart and comment on production after 4 weeks.

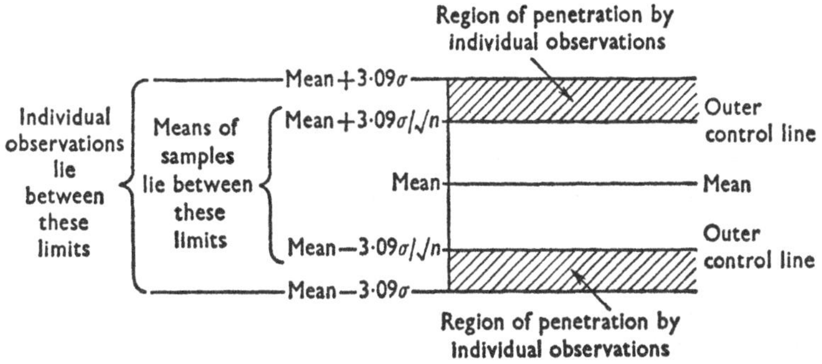

Fig. 15. The shaded regions of width $(3.09\sigma - 3.09\sigma/\sqrt{n})$ are the regions into which individual observations may penetrate even though the means of samples lie between the outer control lines. (We assume here that the practical 100% zone of a normal distribution is 3.09σ either side of the mean.)

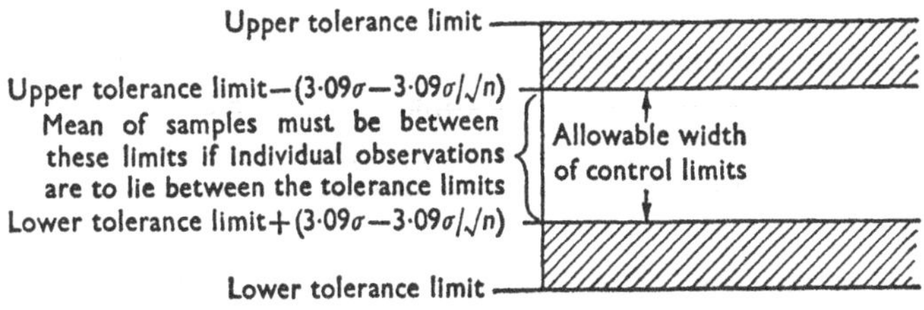

Fig. 16. The allowable width of control limits if the articles produced are to be within the specified tolerance limits.

136. Allowable width of control limits when tolerance limits are specified. When tolerance limits are specified it is necessary to ensure that the articles produced are within the tolerance requirements. Figure 15 illustrates the fact that even when the means of samples are between the outer control lines the individual observations may penetrate into a region of width $(3.09\sigma - 3.09\sigma/\sqrt{n})$ outside the control lines. In fig. 16 this region has been transferred from fig. 15 and placed inside the tolerance limits to give the *allowable width of control limits* within which the means of samples must lie if the individual observations are to be between the tolerance limits.

In the example of §131 let us suppose that the tolerance limits are 0.095 to 1.015. Since $\sigma = 3.787$ thousandths

$$(3.09\sigma - 3.09\sigma/\sqrt{n}) = 5.85 \text{ thousandths}$$

$$= 0.006 \text{ in. (to the nearest thousandth),}$$

and the allowable width of the control limits is

$$0.095 \pm 0.006 \quad \text{to} \quad 1.015 - 0.006,$$

i.e.

$$1.001 \quad \text{to} \quad 1.009.$$

As the means of all 20 samples lie within these limits we conclude that the compression springs produced during the period covered by the sample are within the tolerance requirements.

137. Exercise.

Given that the tolerance limits of the 100-ohm electrical resistors of §133 are 100 ± 9 ohms establish allowable width of control limits based on the first 10 samples and determine whether or not production was to tolerance requirements (i) after one week, (ii) after four weeks.

138. Control chart for fraction defective. It was mentioned in §128 that when articles being manufactured are not classified by dimensions, but are deemed either sound or defective, the quality of production is controlled by an application of the Poisson distribution. It is usually necessary for this type of control to sample at least 20 % of the output. When 20 or more samples have been examined it is possible to set up a control chart which indicates that *production is under control provided not more than one point in ten lies above the control line.* The following example illustrates the method:

A random sample of 20 *articles was drawn from every* 100 *produced by a certain process. The number of defectives per sample in the first* 20 *samples were*

0, 1, 0, 1, 1, 3, 0, 2, 1, 0, 0, 0, 0, 3, 0, 0, 2, 1, 1, 2;

and in the second 20 *samples were*

1, 2, 2, 0, 1, 2, 0, 1, 2, 2, 0, 1, 0, 0, 1, 0, 1, 1, 1, 1.

Use the first 20 *samples to calculate the mean number of defectives per sample and by using the Poisson distribution obtain a* CONTROL LIMIT *c such that the probability of there being c or more defectives in a sample is less than* $\frac{1}{10}$.

Use c to set up a control chart and show that over the whole period during which the samples were drawn the process was under control with a process mean of $4\frac{1}{2}$ % *defective.*

As there were 18 defectives in the first 20 samples the mean number of defectives per sample is 0.9.

By substituting $a = 0.9$ in the Poisson distribution

$$e^{-a}, \quad ae^{-a}, \quad \frac{a^2}{2!}e^{-a}, \quad \frac{a^3}{3!}e^{-a}, \quad \ldots$$

we find the probabilities of 0, 1, 2, 3, ... defectives per sample are

$$0.4066, \quad 0.3659, \quad 0.1647, \quad 0.0494, \quad \ldots.$$

It is now necessary to decide how many of these probabilities must be added together to make a total greater than $\frac{9}{10}$. As the sum of the first three probabilities is 0.9372 the probability of 3 *or more* defectives per sample is less than $\frac{1}{10}$. Thus the required CONTROL LIMIT $c = 3$.

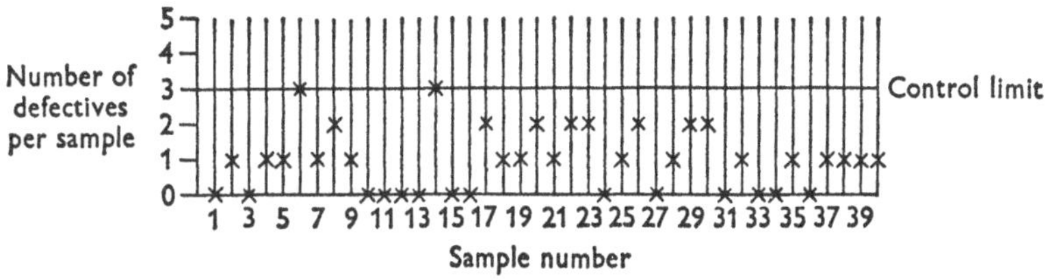

Fig. 17. Quality-control chart for fraction defective. If more than 1 point out of 10 falls above the control limit, the process is out of control.

A control chart can now be set up as shown in fig. 17. As no points fall above the control limit (and only 2 out of 40 fall on it) the process is under control. In the first 20 samples, there are 18 defectives out of a total of 400. This establishes the process mean as $4\frac{1}{2}\%$ and the process is stated to be under control with a process average of $4\frac{1}{2}\%$ defective.

It may be noted that, on the average, there were 9 defectives in every 200 articles tested and when discovered these would be rejected. However, for every 9 defectives discovered and rejected there were 36 undiscovered, because of every 200 articles tested 800 passed by untested. Thus there were finally about 36 defectives remaining in every 991 produced. The application of quality control, therefore, caused the reduction of the process average from $4\frac{1}{2}\%$ defective to 3.6% defective.

In this chapter only a very brief outline of the methods of quality control has been given, but it is sufficient to show how a chart can be used to indicate the overall position of a manufacturing process. It shows up the good results as well as the bad ones. It shows that variability in production, although unavoidable, can be acceptable.

139. Exercise.

A random sample of 20 articles was drawn from every 100 produced by a certain process. The number of defectives per sample in the first 20 samples were

0, 2, 0, 1, 1, 2, 1, 4, 2, 1, 3, 1, 3, 1, 1, 0, 3, 1, 1, 2

and in the second 20 samples were

1, 1, 2, 1, 2, 1, 1, 2, 1, 2, 3, 3, 0, 0, 5, 5, 2, 3, 3, 3.

Use the first 20 samples to calculate the mean number of defectives per sample and then, by the Poisson distribution, obtain the CONTROL LIMIT c such that the probability of there being c or more defectives per sample is less than $\frac{1}{10}$.

Set up a control chart and determine whether or not the process was under control during the whole period.

11

METHOD OF LEAST SQUARES

140. A bivariate distribution. Table 11A shows, side by side, the number of vehicles, x, on the roads of Great Britain and the total casualties, y, in road accidents for the years 1945–54. The source is the *Annual Abstract of Statistics.*

TABLE 11A

	Vehicles with licences current during the September quarter (millions) x	Total casualties in road accidents (thousands) y
1945	2.6	138
1946	3.1	163
1947	3.5	166
1948	3.7	153
1949	4.1	177
1950	4.4	201
1951	4.6	216
1952	4.9	208
1953	5.3	226
1954	5.8	238

The number of vehicles is seen to have steadily increased, and the number of road accidents has also increased though not quite so regularly. The ten (x, y) pairs of values are an example of a *bivariate distribution.*

141. The scatter diagram: direct correlation. If the value x is plotted on a graph against its corresponding value y as shown in fig. 18, a *scatter diagram* is obtained.

Although the points on the scatter diagram do not fall exactly along a straight line they fall within quite a narrow belt. Small values of y correspond to small values of x, large values of y to large values of x, and x and y are said to be *directly correlated.* The exercises at the end of this chapter will provide the student with further examples of *direct correlation.*

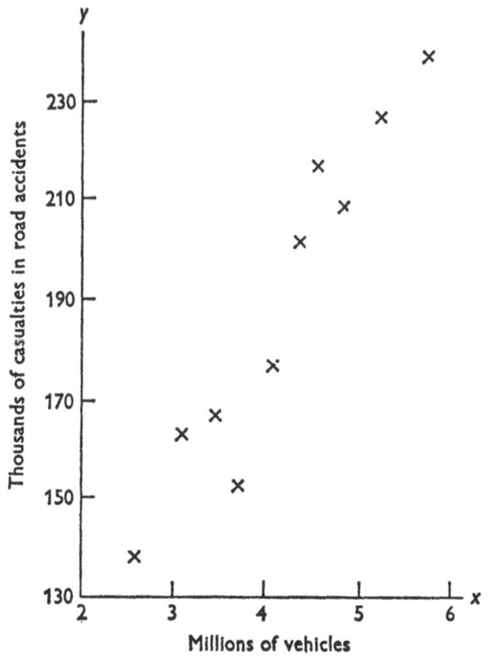

Fig. 18. Scatter diagram showing direct correlation.

142. Inverse correlation. In the bivariate distribution of x and y shown in table 11 B, *small* values of y correspond to *large* values of x and vice versa. This is known as *inverse correlation.* In the scatter diagram (fig. 19) the points lie within a fairly well-defined belt which is downward sloping for increasing values of x.

143. Absence of correlation. Figure 20 is the scatter diagram of the bivariate distribution of y and z of table 11 B. One would not expect correlation in this case because it is unlikely that the *cinema admissions* in one group of towns should be related to the *number of TV licences* issued in a completely different group of towns. This is confirmed by the scatter diagram, the points of which do not lie within any well-defined belt.

144. How to calculate the equation of the least squares line of regression of y on x. Figure 21 shows the scatter diagram of fig. 18 with a straight line fitted through the middle of the ten points. The straight line in this case is *the least squares line of regression of y on x*. It indicates how the *thousands of casualties,* y in any year *depend* on the *millions of vehicles,* x, on the roads in that year. In this case x is the *independent variable* and y the *dependent variable.*

TABLE 11B

Cinema admissions and television licences issued

Year and quarter		Towns served by Sutton Coldfield TV transmitter: Cinema admissions (millions) x	Towns served by Sutton Coldfield TV transmitter: TV licences (per 1000 population) y	Towns not normally served by any TV transmitter: Cinema admissions (millions) z	Towns not normally served by any TV transmitter: TV licences (per 1000 population)
1950	1	11.0	12	12.1	—
	2	10.0	18	11.2	—
	3	10.0	24	11.6	—
	4	9.4	37	10.6	—
1951	1	10.5	52	12.1	—
	2	9.7	64	11.8	—
	3	9.4	69	11.5	—
	4	9.3	81	10.9	1
1952	1	9.9	98	11.6	2
	2	9.3	101	11.1	3
	3	9.0	106	11.3	3
	4	8.6	119	10.5	4

[Northern]

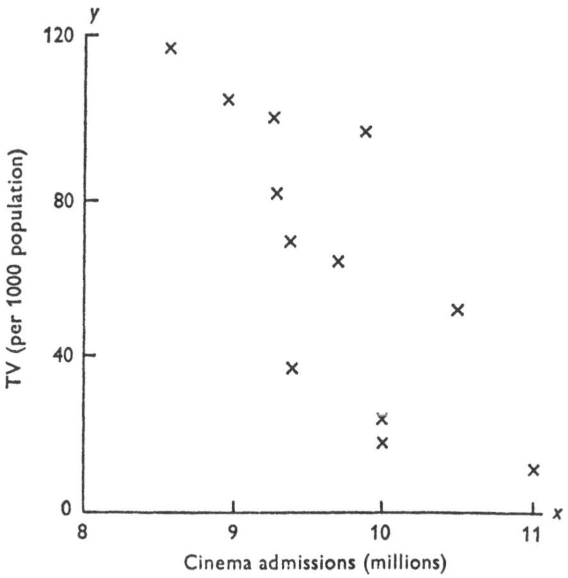

Fig. 19. Scatter diagram showing inverse correlations.

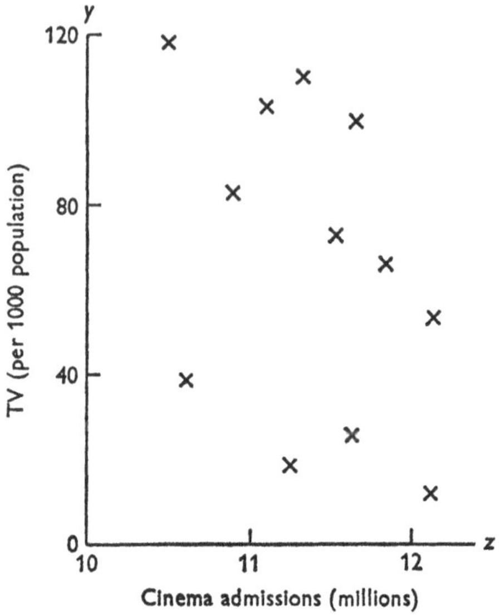

Fig. 20. Scatter diagram of a bivariate distribution in which the variables are not correlated.

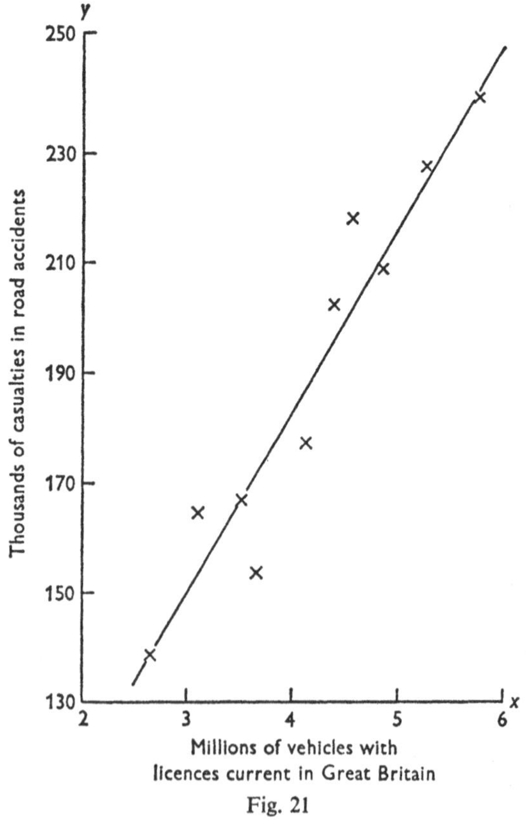

Fig. 21

The equation of the least squares line of regression of y on x for the n points (x_1, y_1), (x_2, y_2), ..., (x_n, y_n) is calculated by first supposing that it is of the form

$$y = a+bx,$$

where a and b are constants. The values of a and b are then found by solving the two simultaneous equations

$$\Sigma y = na+b\Sigma x, \quad (1)$$

$$\Sigma xy = a\Sigma x+b\Sigma x^2, \quad (2)$$

where $\Sigma x = x_1+x_2+...+x_n$, $\Sigma y = y_1+y_2+...+y_n$,

$$\Sigma x^2 = x_1^2+x_2^2+...+x_n^2, \quad \Sigma xy = x_1y_1+x_2y_2+...+x_ny_n.$$

The equations (1) and (2) above are called the *normal equations*. Their formal derivation is given in §145. Note that equation (1) is obtained from $y = a+bx$ by placing Σ before y and x and n before a, and equation (2) is obtained by first multiplying $y = a+bx$ by x and then placing Σ before xy, x^2 and x. The actual calculation of the equation of the straight line shown in fig. 21 is given in table 11c.

TABLE 11c

Calculation of the equation of the least squares lines of regression of y on x

x	y	x^2	xy
2.6	138	6.76	358.8
3.1	163	9.61	505.3
3.5	166	12.25	581.0
3.7	153	13.69	566.1
4.1	177	16.81	725.7
4.4	201	19.36	884.4
4.6	216	21.16	993.6
4.9	208	24.01	1019.2
5.3	226	28.09	1197.8
5.8	238	33.64	1380.4
$\Sigma x = 42.0$	$\Sigma y = 1886$	$\Sigma x^2 = 185.38$	$\Sigma xy = 8212.3$

The constants a and b of the straight line $y = a+bx$ are given by the equations

$$1886 = 10a+42.0b,$$

$$8212.3 = 42.0a+185.38b.$$

Thus $a = 53$, $b = 32$ and the equation of the least squares line of regression of y on x is

$$y = 53+32x.$$

145. The formal derivation of the normal equations. Suppose that $P_1, P_2, \ldots, P_n$ are n points whose co-ordinates are

$$(x_1, y_1), \quad (x_2, y_2), \quad \ldots, \quad (x_n, y_n)$$

respectively and that AB is a straight line whose equation is $y = a+bx$ (see fig. 22). The ordinates $P_1R_1, P_2R_2, \ldots, P_nR_n$ have respective lengths $y_1, y_2, \ldots, y_n$ and if $Q_1, Q_2, \ldots, Q_n$ are the respective points of intersection of the ordinates with the line AB then $Q_1R_1, Q_2R_2, \ldots, Q_nR_n$ have

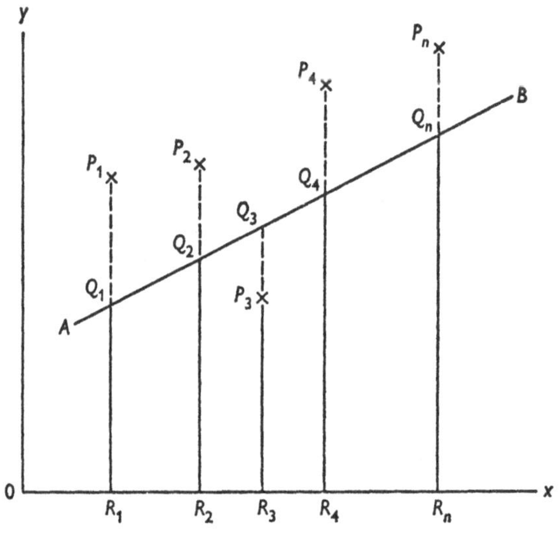

Fig. 22. The least squares line of regression of y on x is such that the sum of the squares of the residuals $P_1Q_1^2+P_2Q_2^2+\ldots+P_nQ_n^2$ is a minimum.

respective lengths $(a+bx_1), (a+bx_2), \ldots, (a+bx_n)$. Thus the *residuals* $P_1Q_1, P_2Q_2, \ldots, P_nQ_n$ have respective lengths

$$(y_1-a-bx_1), \quad (y_2-a-bx_2), \quad \ldots, \quad (y_n-a-bx_n),$$

and if z represents the sum of the squares of the lengths of the residuals

$$z = (y_1-a-bx_1)^2+(y_2-a-bx_2)^2+\ldots+(y_n-a-bx_n)^2.$$

By varying the constants a and b the position of the line AB can be altered. If a varies while b remains constant the line takes up a series of positions keeping a constant direction. If b varies while a remains constant the line rotates about a fixed point on the y-axis.

The least squares line of regression of y on x is the line whose position is chosen so that z is a minimum. Differentiating z with respect to a and regarding b as constant we obtain

$$\frac{\partial z}{\partial a} = -2(y_1-a-bx_1)-2(y_2-a-bx_2)-\ldots-2(y_n-a-bx_n),$$

and differentiating z with respect to b and regarding a as constant we obtain

$$\frac{\partial z}{\partial b} = -2x_1(y_1-a-bx_1)-2x_2(y_2-a-bx_2)-\ldots-2x_n(y_n-a-bx_n).$$

If z is a minimum $\quad \frac{\partial z}{\partial a} = 0$ and $\frac{\partial z}{\partial b} = 0.$

Thus $\quad (y_1-a-bx_1)+(y_2-a-bx_2)+\ldots+(y_n-a-bx_n) = 0$

and $\quad x_1(y_1-a-bx_1)+x_2(y_2-a-bx_2)+\ldots+x_n(y_n-a-bx_n) = 0.$

Using the Σ notation these equations can be expressed as

$$\Sigma y - na - b\Sigma x = 0$$

and

$$\Sigma xy - a\Sigma x - b\Sigma x^2 = 0.$$

Thus, if $y = a+bx$ is the equation of the least squares line of regression of y on x, the values of a and b are given by the normal equations

$$\Sigma y = na+b\Sigma x,$$

$$\Sigma xy = a\Sigma x+b\Sigma x^2.$$

146. The regression line passes through the mean of the array. The first of the normal equations $\Sigma y = na+b\Sigma x$ can be written in the form

$$\frac{\Sigma y}{n} = a+b\frac{\Sigma x}{n}$$

or

$$\bar{y} = a+b\bar{x},$$

where $\bar{y}$ represents the mean of the n values of y and $\bar{x}$ represents the mean of the n values of x. Thus the co-ordinates $(\bar{x}, \bar{y})$ of the point known as the mean of the array of points (x_1, y_1), (x_2, y_2), ..., (x_n, y_n) satisfy the equation $y = a+bx$. This implies that the regression line passes through the mean of the array.

Reference to table 11C shows that the mean of the array of points in fig. 21 is (4.2, 188.6) and the regression line will be observed to pass through it.

147. The regression coefficient. The constant b which is the slope of the regression line is known as the *coefficient of regression of y on x*. Table 11C indicates that the regression coefficient of that particular example is 32 thousand casualties per million vehicles.

The student will find it an easy exercise to deduce from the normal equations that the general value of b is given by

$$b = \frac{n\Sigma xy - \Sigma x\Sigma y}{n\Sigma x^2 - (\Sigma x)^2}$$

$$= \frac{\dfrac{\Sigma xy}{n} - \left(\dfrac{\Sigma x}{n}\right)\left(\dfrac{\Sigma y}{n}\right)}{\dfrac{\Sigma x^2}{n} - \left(\dfrac{\Sigma x}{n}\right)^2}.$$

Now the *variance* of $x_1, x_2, \ldots, x_n$ is defined as

$$s_x^2 = \frac{\Sigma(x-\bar{x})^2}{n}$$

$$= \frac{\Sigma x^2 - 2\bar{x}\Sigma x + n\bar{x}^2}{n}$$

$$= \frac{\Sigma x^2}{n} - \left(\frac{\Sigma x}{n}\right)^2$$

and the *covariance* of $(x_1, y_1), (x_2, y_2), \ldots, (x_n, y_n)$ is defined as

$$s_{xy} = \frac{\Sigma(x-\bar{x})(y-\bar{y})}{n}$$

$$= \frac{\Sigma xy - \bar{x}\Sigma y - \bar{y}\Sigma x + n\bar{x}\bar{y}}{n}$$

$$= \frac{\Sigma xy}{n} - \left(\frac{\Sigma x}{n}\right)\left(\frac{\Sigma y}{n}\right).$$

Thus the coefficient of regression $b = s_{xy}/s_x^2$ and since the regression line passes through $(\bar{x}, \bar{y})$ an alternative form of its equation is

$$(y-\bar{y}) = \frac{s_{xy}}{s_x^2}(x-\bar{x}).$$

This formula is used in my *First Course in Statistics* to fix the position of the regression line.

148. The line of regression of x on y. If values of x are to be estimated from known values of y the line of regression of x on y is used. This line has an equation

$$x = a' + b'y,$$

where a' and b' are found from the normal equations

$$\Sigma x = na' + b'\Sigma y$$

and

$$\Sigma xy = a'\Sigma y + b'\Sigma y^2.$$

The value of b' in this case is the *coefficient of regression of x on y*. By interchanging x and y in §147 we find that

$$b' = \frac{n\Sigma xy - \Sigma x \Sigma y}{n\Sigma y^2 - (\Sigma y)^2}$$

$$= \frac{s_{xy}}{s_y^2}$$

and an alternative form of the equation of the line of regression of x on y is

$$(x-\bar{x}) = \frac{s_{xy}}{s_y^2}(y-\bar{y}).$$

This is the case in which y is the *independent variable* and x the *dependent variable*. The line of regression of x on y is illustrated by fig. 23 in which the residuals are parallel to the x-axis.

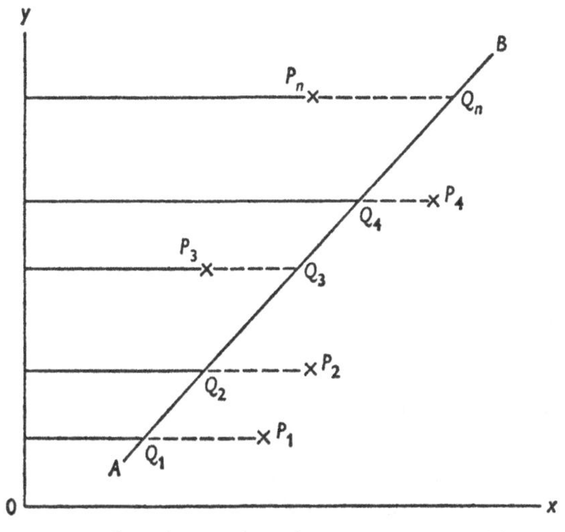

Fig. 23. The least squares line of regression of x on y, represented by AB above, is such that the sum of the squares of the residuals

$$P_1Q_1^2 + P_2Q_2^2 + \ldots + P_nQ_n^2$$

is a minimum.

149. The two lines of regression in one diagram. Figure 24 is fig. 19 with the two regression lines fitted. AB is the line of regression of x on y, while CD is the line of regression of y on x. AB is used to estimate the cinema admissions for a given number of TV licences per 1000 population. CD would be used to estimate the TV licences per 1000 population for a given number of cinema admissions. As the effect of the new form of entertainment on the old is of greater interest than the effect of the old on the new, AB is of greater use than CD.

150. Dichotomy. If the two variables of a bivariate distribution are very closely correlated the points of the scatter diagram will lie within a very narrow belt and the angle between the two regression lines will be very small. For more widely dispersed points the angle between the regression lines is greater and the correlation is said to be less marked.

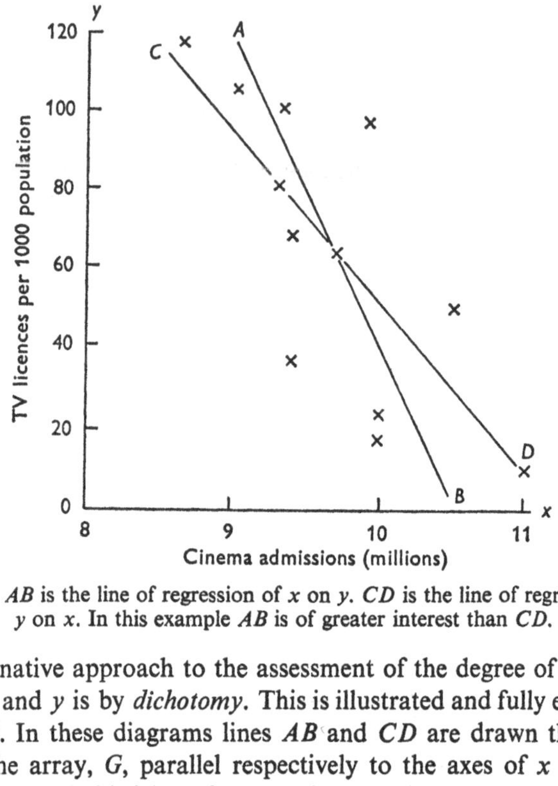

Fig. 24. *AB* is the line of regression of *x* on *y*. *CD* is the line of regression of *y* on *x*. In this example *AB* is of greater interest than *CD*.

An alternative approach to the assessment of the degree of correlation between *x* and *y* is by *dichotomy.* This is illustrated and fully explained in figs. 25–29. In these diagrams lines *AB* and *CD* are drawn through the, mean of the array, *G*, parallel respectively to the axes of *x* and *y*. The points are thus divided into four quadrants. The values of *x* are *dichotomized* by *CD*, that is to say, divided into two classes. The values of *y* are dichotomized by *AB* and this process is called dichotomy.

151. Exercises.

1. The following table gives the percentage of sand in soil at different depths:

										Total
x (depth in mm)	0	150	300	450	600	750	900	1050	1200	5400
y (% sand)	80.6	63.0	64.3	62.5	57.5	59.2	40.8	46.9	37.6	512.4

and the following sums of squares and products were calculated from the above data:

$$\Sigma(x-\bar{x})^2 = 1\,350\,000,$$
$$\Sigma(x-\bar{x})(y-\bar{y}) = -40\,590,$$
$$\Sigma(y-\bar{y})^2 = 1422.36.$$

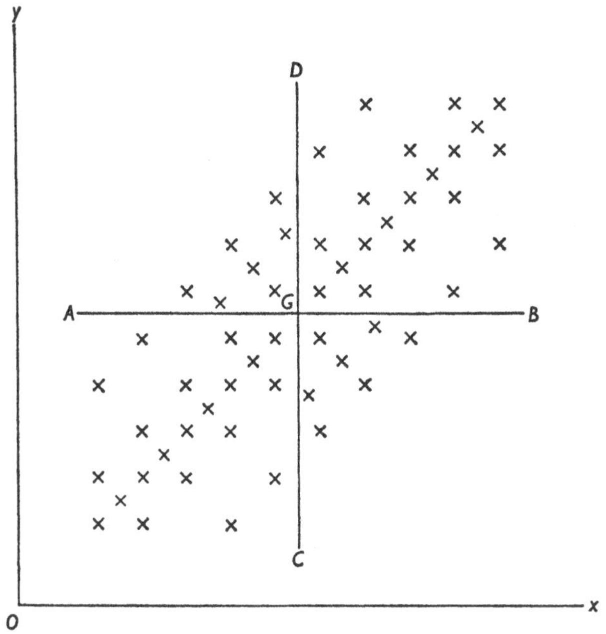

Fig. 25. Direct (or positive) correlation indicated by dichotomy. The lines *AB* and *CD* are drawn through the mean of the array, *G*, parallel respectively to the axes *x* and *y*. In this diagram a *high proportion* of the points lie in the quadrants *BGD* and *AGC*. This is an indication that *x* and *y* are directly correlated.

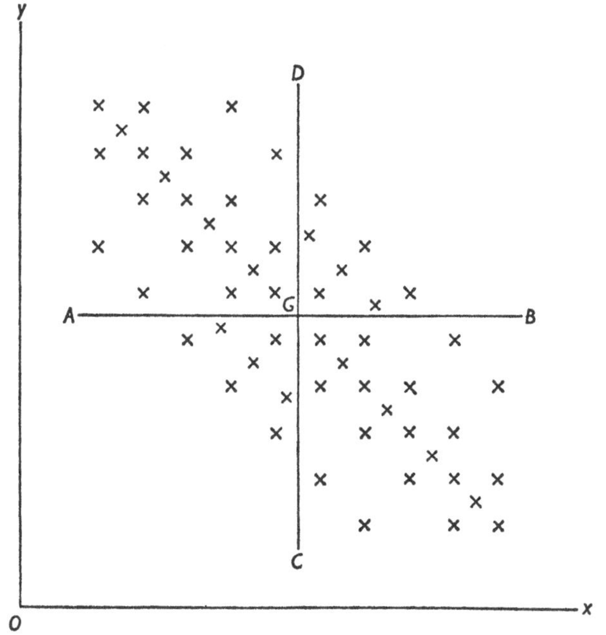

Fig. 26. Inverse (or negative) correlation indicated by dichotomy. In this case only a *small proportion* of the points lie in the quadrants *BGD* and *AGC*. This indicates that *x* and *y* are inversely correlated.

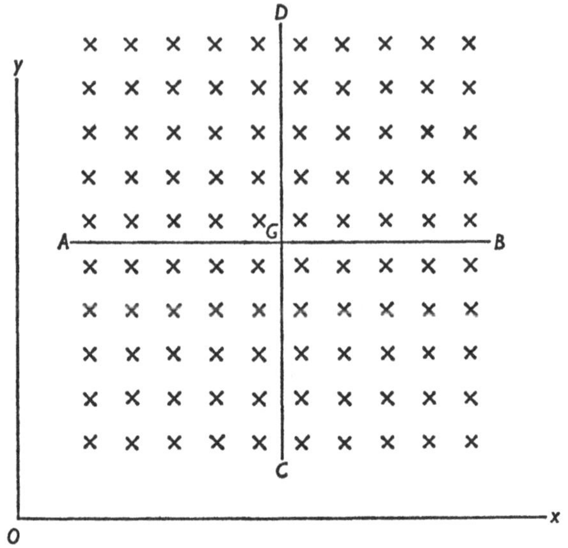

Fig. 27. Complete absence of correlation. Equal proportions of the points in all four quadrants.

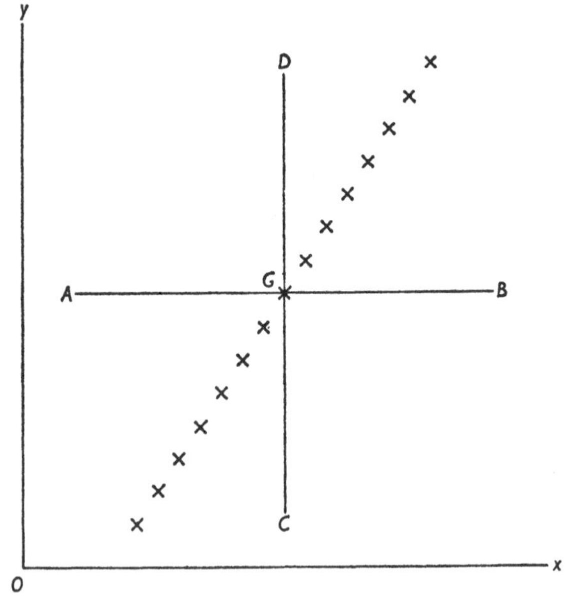

Fig. 28. Perfect direct correlation. All the points lie on a straight line.

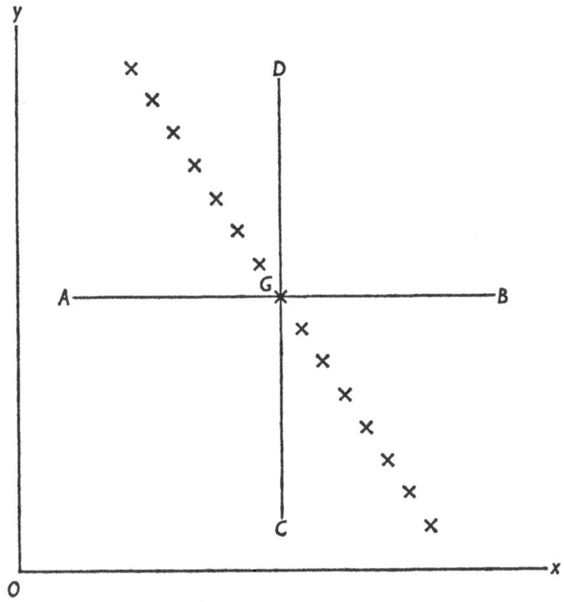

Fig. 29. Perfect inverse correlation.

Find (giving the numbers correct to three significant figures) the equations of the regression lines which you would use (i) to predict the depth from the percentage sand, (ii) to estimate the dependence of percentage of sand on depth.

Show these lines (labelled (i) and (ii)) on a scatter diagram of the original data. [London]

2. *Percentage shrinkage in samples of cloth after washing, in directions along and across the cloth*

Along (x)	Across (y)	Along (x)	Across (y)
12	5	7	5
4	2	12	7
10	5	18	10
10	8	14	7
11	6	14	8
10	8	8	4
6	3	11	6
6	4	17	8
6	3	21	11
13	5	12	9

Calculate the equation of the line of regression of x on y.

A roll of cloth is sampled by cutting a narrow test strip right across the roll. The strip proves to have a percentage shrinkage of 7. Use your regression equation to obtain an estimate of the percentage shrinkage to be expected along the cloth.

It is desired to cut from the roll a piece of cloth which may be expected to shrink to 10 in. square after washing. Describe how this piece should be cut. [Northern]

3. Five groups of locusts, each containing 120, were exposed to a lethal spray in various concentrations. The deaths resulting were as follows:

Concentration (multiple of standard)	1.2	1.4	1.6	1.8	2.0
No. of deaths	38	52	46	76	66

For each concentration, find the percentage of locusts dying. Plot this percentage against the concentration on a scatter diagram and calculate the equation of the line of regression, taking concentration as your independent variable. [Northern]

4. To find the electrical resistance r of a wire, it was connected in a Wheatstone bridge circuit and the following series of readings were made:

x	0.1	0.2	0.3	0.4	0.5	0.6	0.7	0.8	0.9	1.0
y	0.270	0.476	0.702	0.910	1.106	1.350	1.560	1.742	1.994	2.182

Assuming that errors occur in the values of y but not in the values of x and that x and y satisfy the equation $y = rx + R$ where R is the resistance of the rest of the circuit, find the best values of r and R.

5. The number of grams of a given salt which will dissolve in 100 g of water at different temperatures is shown in the table below:

Temperature (x °C)	0	10	20	30	40	50	60	70	80	90	100
Weight of salt (y g)	53.5	59.5	65.2	70.6	75.5	80.2	85.5	90.0	95.0	99.2	104.0

Use the method of least squares to find the linear formula $y = a + bx$ which best fits these observations. [London]

6. A general knowledge test consisting of a hundred questions was given to fifteen boys of different ages with results as follows:

Boy	Age: Years	Age: Months	No. of questions correct
A	11	7	18
B	11	1	19
C	12	8	23
D	12	0	26
E	13	5	25
F	13	6	31
G	14	9	24
H	15	3	32
I	14	7	28
J	15	6	25
K	15	9	33
L	15	7	31
M	16	11	36
N	17	1	32
O	16	10	40

Plot a scatter diagram of the numbers of questions correct (y) against the age in months (x), using 1 in. to represent 10 correct questions on the y-axis and 1 in. to represent 10 months on the x-axis. Calculate the equation of the line of regression of y on x and show the line on the diagram.

State, with reasons, which boy deserves the prize for the best performance taking age into consideration. [Northern]

7. To fit a parabola $y = a+b_1x+b_2x^2$ to data by the method of least squares we find the values of a, b, c that will make z a minimum where

$$z = (y_1-a-b_1x_1-b_2x_1^2)^2+(y_2-a-b_1x_2-b_2x_2^2)^2+\ldots+(y^n-a-b_1x^n-b_2x_n^2)^2.$$

By letting $$\frac{\partial z}{\partial a} = 0, \quad \frac{\partial z}{\partial b_1} = 0 \quad \text{and} \quad \frac{\partial z}{\partial b_2} = 0$$

show that the *normal* equations are

$$\Sigma y = na+b_1\Sigma x+b_2\Sigma x^2,$$

$$\Sigma xy = a\Sigma x+b_1\Sigma x^2+b_2\Sigma x^3,$$

$$\Sigma x^2y = a\Sigma x^2+b_1\Sigma x^3+b_2\Sigma x^4.$$

8. *Result of fertiliser experiment on crop results*

Units of fertiliser used	0	2	4	6	8	10
Units of yield	110	113	118	119	120	118

Fit a parabola (see Ex.7 above) to the above data, and estimate for what fertiliser application the best results are obtained. [London]

12

CORRELATION BY PRODUCT MOMENTS

152. The coefficient of correlation r_{xy}. An elementary treatment of the coefficient of correlation was given in my *First Course in Statistics*. The following introduction is more formal.

The line of regression of y on x was established in §147 as

$$(y-\bar{y}) = \frac{s_{xy}}{s_x^2}(x-\bar{x}).$$

If this equation is *standardised* by putting it in the form

$$\frac{(y-\bar{y})}{s_y} = \frac{s_{xy}}{s_x s_y}\frac{(x-\bar{x})}{s_x}$$

it can be written

$$Y = r_{xy}X,$$

where the symbol r_{xy} represents $s_{xy}/s_x s_y$ and the axes of X and Y have their origin at the mean of the array, the X-axis being graduated in units equal to s_x and the Y-axis in units equal to s_y.

Similarly the line of regression of X on Y can be put in the form $X = r_{xy}Y$.

The two lines of regression referred to their new axes will then appear as shown in fig. 30. It will be seen that $Y = r_{xy}X$ makes an angle θ with the X-axis where $\tan\theta = r_{xy}$ and $X = r_{xy}Y$ makes the same angle θ with the Y-axis.

If the correlation is direct and the points of the scatter diagram lie exactly along a straight line, the two lines of regression coincide, θ is 45° and $r_{xy} = 1$.

If the correlation is direct and the points of the scatter diagram are dispersed on either side of a straight line, the lines of regression are separate as shown in fig. 30. The case of greatest dispersion is when $\theta = 0$ and $r_{xy} = 0$. In this case the two regression lines coincide with the axes of X and Y.

If the correlation is inverse, θ is negative and the two lines of regression lie in the second and fourth quadrants instead of the first and third.

If the correlation is inverse and the points of the scatter diagram lie exactly on a straight line, the two regression lines coincide, $\theta = -45°$ and $r_{xy} = -1$.

We define $r_{xy} = s_{xy}/(s_x s_y)$ as the *coefficient of correlation.* Its values lie between $+1$ and -1 and are positive for direct correlation and negative for inverse correlation. A coefficient of $+1$ indicates perfect direct correlation, a coefficient of -1 perfect inverse correlation and a coefficient of 0 indicates complete absence of correlation.

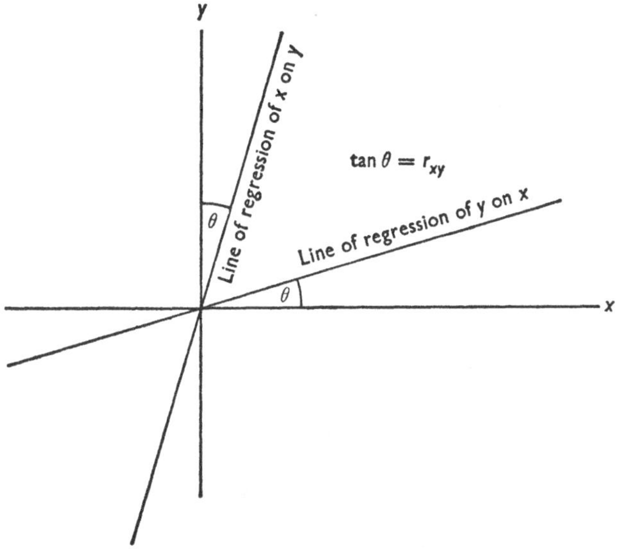

Fig. 30. The two regression lines when the units have been standardized and the mean of the array has been made the origin of co-ordinates.

153. The calculation of r_{xy}. The numerical value of r_{xy} can be calculated by the formula $r_{xy} = s_{xy}/(s_x s_y)$ where the covariance

$$s_{xy} = \frac{1}{n}\,\Sigma(x-\bar{x})(y-\bar{y}) \qquad \text{(see §147)}$$

$$= \text{the mean product of deviations,}$$

the standard deviation of the n values of x

$$s_x = \sqrt{\left\{\frac{\Sigma(x-\bar{x})^2}{n}\right\}}$$

and the standard deviation of the n values of y

$$s_y = \sqrt{\left\{\frac{\Sigma(y-\bar{y})^2}{n}\right\}}.$$

This procedure was used in *A First Course in Statistics.*

TABLE 12A

The calculation of the coefficient of correlation r_{xy} (working with an arbitrary origin and convenient units)

Millions of vehicles (working with 4.1 as origin and 0.1 as unit) x	Thousands of casualties (working with 201 as origin) y	x^2 (0.01 as unit)	y^2	xy (0.1 as unit)
−15	−63	225	3969	945
−10	−38	100	1444	380
−6	−35	36	1225	210
−4	−48	16	2304	192
0	−24	0	576	0
3	0	9	0	0
5	15	25	225	75
8	7	64	49	56
12	25	144	625	300
17	37	289	1369	629
$\Sigma x = 10$	$\Sigma y = -128$	$\Sigma x^2 = 908$	$\Sigma y^2 = 11786$	$\Sigma xy = 2787$

Covariance $$s_{xy} = \frac{\Sigma xy}{n} - \left(\frac{\Sigma x}{n}\right)\left(\frac{\Sigma y}{n}\right)$$

$= 278.7-(1)(-12.8)$ with 0.1 as unit

$= 29.15$ in the original units.

Standard deviation of the 10 values of x

$$s_x = \sqrt{\left\{\frac{\Sigma x^2}{n} - \left(\frac{\Sigma x}{n}\right)^2\right\}}$$

$= \sqrt{\{90.8-(1)^2\}}$ with 0.1 as unit

$= 0.9476$ in the original units.

Standard deviation of the 10 values of y

$$s_y = \sqrt{\left\{\frac{\Sigma y^2}{n} - \left(\frac{\Sigma y}{n}\right)^2\right\}}$$

$= \sqrt{\{1178.6-(-12.8)^2\}}$

$= 31.86$ in the original units.

Coefficient of correlation $r_{xy} = s_{xy}/(s_x.s_y)$

$= 0.966,$

The more advanced student may prefer, however, to use

$$s_{xy} = \frac{\Sigma xy}{n} - \left(\frac{\Sigma x}{n}\right)\left(\frac{\Sigma y}{n}\right),$$

$$s_x = \sqrt{\left\{\frac{\Sigma x^2}{n} - \left(\frac{\Sigma x}{n}\right)^2\right\}},$$

$$s_y = \sqrt{\left\{\frac{\Sigma y^2}{n} - \left(\frac{\Sigma y}{n}\right)^2\right\}},$$

changing, if necessary, the origin and the units in order to minimize the amount of arithmetic. Table 12A shows the calculation of r_{xy} by this second method for the bivariate distribution given in § 140. Note that the xy values are not necessarily positive.

154. The coefficient of correlation and the regression lines for grouped data. A survey was made of the weights and heights of 248 boys between the ages of 11 and 16 years. Table 12B is a classification of the results of the survey. The number 28 in the table indicates, for example, that there were

TABLE 12B

Height of boy	Weight of boy in kilos									
in cm	25–	31–	37–	43–	49–	55–	61–	67–	73–	79–
130.0–	—	1	—	—	—	—	—	—	—	—
137.5–	4	12	4	—	—	—	—	—	—	—
145.0–	—	22	21	3	—	—	—	—	—	—
152.5–	—	6	28	14	1	1	—	—	—	—
160.0–	—	1	1	24	17	4	—	—	1	—
167.5–	—	—	—	4	22	18	2	3	—	—
175.0–	—	—	—	—	1	14	10	4	—	—
182.5–	—	—	—	—	—	2	2	1	—	—
190.0–	—	—	—	—	—	—	—	—	—	—

28 boys of heights between 152.5 and 160.0 cm whose weights were between 37 and 43 kilos. The calculation of r_{xy} from *grouped data* such as this is shown in table 12C in which

(i) the central number in each *cell* is the frequency f of the (x, y) pair of values of the cell,

(ii) the upper left-hand italic number in each cell is the product xy of the (x, y) pair of values of the cell,

(iii) the lower right-hand italic number in each cell is the *product moment* fxy of the cell found by multiplying the central number by the upper left-hand number.

TABLE 12C

The calculation of the coefficient of correlation r_{xy} from grouped data

		Weight of boy 46 kilos as origin and 6 kilos as unit (x)									Horizontal totals for each value of y f_y	First moment of each value of y $f_y y$	Second moment of each value of y $f_y y^2$
		−3	−2	−1	0	1	2	3	4	5			
Height of boy 56.25 cm as origin and 7.5 cm as unit	−3		*6* 1 *6*								1	−3	9
	−2	*6* 4 *24*	*4* 12 *48*	*2* 4 *8*							20	−40	80
	−1		*2* 22 *44*	*1* 21 *21*	*0* 3 *0*						46	−46	46
	0		*0* 6 *0*	*0* 28 *0*	*0* 14 *0*	*0* 1 *0*	*0* 1 *0*				50	0	0
(y)	1		*−2* 1 *−2*	*−1* 1 *−1*	*0* 24 *0*	*1* 17 *17*	*2* 4 *8*			*5* 1 *5*	48	48	48
	2				*0* 4 *0*	*2* 22 *44*	*4* 18 *72*	*6* 2 *12*	*8* 3 *24*		49	98	196
	3					*3* 1 *3*	*6* 14 *84*	*9* 10 *90*	*12* 4 *48*		29	87	261
	4						*8* 2 *16*	*12* 2 *24*	*16* 1 *16*		5	20	80
Vertical total for each value of x	f_x	4	42	54	45	41	39	14	8	1	$248 = \Sigma f$	$164 = \Sigma f_y y$	$720 = \Sigma f_y y^2$
First moment of each value of x	$f_x x$	−12	−84	−54	0	41	78	42	32	5	$48 = \Sigma f_x x$		
Second moment of each value of x	$f_x x^2$	36	168	54	0	41	156	126	128	25	$734 = \Sigma f_x x^2$		
Vertical total of product moments		*24*	*96*	*28*	*0*	*64*	*180*	*126*	*88*	*5*	$611 = \Sigma fxy$		

By table 12C, therefore,

the mean weight $\bar{x} = \Sigma f_x x \Sigma f$

$= 48/248$ 6 kg units with 46 kg as origin

$= 47.16$ kg;

the mean height $\bar{y} = \Sigma f_y y/\Sigma f$

$= 164/248$ 7.5 cm units with 156.25 cm as origin

$= 161.2$ cm;

the standard deviation of the x array

$$s_x = \sqrt{\left\{\frac{\Sigma f_x x^2}{\Sigma f} - \left(\frac{\Sigma f_x x}{\Sigma f}\right)^2\right\}}$$

$$= \sqrt{\left\{\frac{734}{248} - \left(\frac{48}{248}\right)^2\right\}} \quad \text{6 kg units}$$

$$= 10.26 \text{ kg};$$

the standard deviation of the y array

$$s_y = \sqrt{\left\{\frac{\Sigma f_y y^2}{\Sigma f} - \left(\frac{\Sigma f_y y}{\Sigma f}\right)^2\right\}}$$

$$= \sqrt{\left\{\frac{720}{248} - \left(\frac{164}{248}\right)^2\right\}} \quad \text{7.5 cm units}$$

$$= 11.78 \text{ cm};$$

the covariance
$$s_{xy} = \frac{\Sigma fxy}{\Sigma f} - \left(\frac{\Sigma f_x x}{\Sigma f}\right)\left(\frac{\Sigma f_y y}{\Sigma f}\right)$$

$$= \frac{611}{248} - \left(\frac{48}{248}\right)\left(\frac{164}{248}\right) \quad \text{6 kg} \times \text{7.5 cm units}$$

$$= 105.1 \text{ kg} \times \text{cm units.}$$

Hence the *product moment coefficient of correlation*

$$r_{xy} = s_{xy}/(s_x . s_y)$$

$$= 105.1/(10.26 \times 11.78)$$

$$= 0.87$$

and the least squares line of regression of x on y is

$$(x - \bar{x}) = \frac{s_{xy}}{s_y^2}(y - \bar{y}),$$

that is
$$x - 47.16 = \frac{105.1}{138.7}(y - 161.2)$$

which may be written $x = 0.76y - 75.$

Note that the line of regression of x on y is of greater interest, in this case, than the line of regression of y on x. It enables us to state the average weight of a boy of given height. This is useful because a deviation from the average weight (for a given height) might be a help in assessing the state of health.

155. The significance of r_{xy}. It was indicated in §152 that the coefficient of correlation r_{xy} lies between $+1$ and -1 and is positive for direct correlation and negative for inverse correlation. Values of r_{xy} near to unity indicate a high degree of correlation, values near to zero indicate an absence of correlation. When the value of r_{xy} has been calculated the question therefore arises 'Does $|r_{xy}|$ differ significantly from zero?' Table 12D gives the $P = 5\%$ values of $|r_{xy}|$. If the calculated value of $|r_{xy}|$ is equal to or greater than the value given in the table for the appropriate number of (x, y) pairs, $|r_{xy}|$ differs significantly from zero. It differs from zero at the 5 % level of significance. This means that the probability that no association exists between the variables is $\frac{1}{20}$ or less. Finally, note that correlation should not be used at all if either variable is selected (i.e. does not arise from random sampling) though regression of the random variable on the selected variable may still be used. Also r_{xy} is not meaningful if the regression graphs are not straight lines.

TABLE 12D

The $P = 5\%$ values of $|r_{xy}|$

No. of pairs of values of x and y from which r_{xy} is calculated	Minimum value of $\vert r_{xy}\vert$ for correlation to be probable	No. of pairs of values of x and y from which r_{xy} is calculated	Minimum value of $\vert r_{xy}\vert$ for correlation to be probable
5	0.88	14	0.54
6	0.82	15	0.52
7	0.76	16	0.50
8	0.71	18	0.47
9	0.67	20	0.45
10	0.64	40	0.31
11	0.61	80	0.22
12	0.58	100	0.20
13	0.56		

156. Exercises.

1. Calculate r_{xy} for the bivariate distribution of §142.

2–5. Calculate the product-moment coefficients of correlation for the data given in Exs. 1, 2, 3 and 6, §151.

6. The 1 % sample of the 1951 Census shows the ages of husband and wife to be related as shown below:

Age of wife (y)	Age of husband (x): 20–	30–	40–	50–	60–	70–80
20–	12	7	—	—	—	—
30–	1	18	8	1	—	—
40–	—	2	19	6	1	—
50–	—	—	2	14	5	—
60–	—	—	—	1	8	2
70–80	—	—	—	—	1	3

(1 unit = 100000 couples.)

Calculate the coefficient of correlation r_{xy} and also the equation of the least squares line of regression of y on x.

Estimate the mean age of wife for husbands aged (i) 25, (ii) 45, (iii) 65.

7. *Associated TeleVision Limited. Charges in £ for 15-sec advertisements*

Charge (£y) for a 15-sec advertisement	Number (x) of homes viewing ATV programmes, in thousands: 100–	200–	300–	400–	500–	600–	700–	800–
0–	6	—	—	—	—	—	—	—
50–	2	5	2	—	—	—	—	—
100–	5	7	3	1	1	—	—	—
150–	2	6	1	3	1	2	—	—
200–	1	17	7	0	1	0	—	—
250–	—	—	—	2	1	1	—	—
300–	—	—	—	—	1	1	—	—
350–	—	—	—	—	1	0	—	—
400–	—	—	—	—	—	3	1	—
450–	—	—	—	—	—	—	—	—

The above table shows how the charges for advertisements, £y, made by ATV were related to the number of homes viewing, x, during the autumn of 1956. Thus the number 17 in the table indicates that there were 17 cases in which the charge was between £200 and £250 when between 200000 and 300000 homes were viewing.

Calculate the coefficient of correlation r_{xy} and also the equation of the least squares line of regression of y on x.

Estimate the mean charge made by ATV for a 15-sec advertisement at a time when half-a-million homes were viewing. [Northern]

13

CORRELATION BY RANKS

157. The coefficient of rank correlation. Table 13A shows the data of §140 with *ranks* attached to the x and y values:

TABLE 13A

No. of vehicles (millions) x	Rank X of the x values	No. of casualties (thousands) y	Rank Y of the y values
2.6	10	138	10
3.1	9	163	8
3.5	8	166	7
3.7	7	153	9
4.1	6	177	6
4.4	5	201	5
4.6	4	216	3
4.9	3	208	4
5.3	2	226	2
5.8	1	238	1

The number 2.6, for example, is *tenth* in *order* or *rank* of the x values, 4.1 is *sixth* and 5.8 is *first*. If the product moment coefficient or correlation is calculated from the ranks X and Y instead of the original values of x and y far less arithmetic is involved and an approximation to r_{xy} is obtained which is called the *coefficient of rank correlation*, ρ. This idea was originally introduced by C. Spearman in 1906. By applying the method shown in table 12A to the X and Y values the student will obtain $\rho = 0.95$.

158. The formula for the coefficient of rank correlation. The ranks are the numbers 1, 2, 3, 4, ..., n (n in the example under consideration being 10), and the algebraic formulae for the sum of the first n natural numbers and for the sum of their squares can be used to prove that ρ may be calculated as shown in table 13B by the formula

$$\bar{y} = 1 - \frac{6\Sigma D^2}{n(n^2-1)},$$

where D is the *rank difference*.

TABLE 13B

Calculation of the coefficient of rank correlation (direct)

Rank of no. of vehicles X	Rank of no. of casualties Y	Rank difference $X-Y$ D	D^2
10	10	0	0
9	8	1	1
8	7	1	1
7	9	−2	4
6	6	0	0
5	5	0	0
4	3	1	1
3	4	−1	1
2	2	0	0
1	1	0	0
	Total	0	8

Since $n = 10$ and $\Sigma D^2 = 8$,

$$\rho = 1-6\times 8/10(10^2-1)$$
$$= 0.95.$$

159. The derivation of the formula for the coefficient of rank correlation. Suppose that $X_1, X_2, \ldots, X_n$ are the respective *ranks* of the n observations $x_1, x_2, \ldots, x_n$. Then $X_1, X_2, \ldots, X_n$ are the first n natural numbers $1, 2, \ldots, n$ (though not in order). Similarly if $Y_1, Y_2, \ldots, Y_n$ are the respective ranks of the n observations $y_1, y_2, \ldots, y_n$ then $Y_1, Y_2, \ldots, Y_n$ are the numbers $1, 2, \ldots, n$ (again not in order). It is proved in textbooks of algebra that

$$1+2+3+\ldots+n = \tfrac{1}{2}n(n+1)$$

and
$$1^2+2^2+3^2+\ldots+n^2 = \tfrac{1}{6}n(n+1)(2n+1).$$

Thus
$$\Sigma X = \tfrac{1}{2}n(n+1),$$
$$\Sigma X^2 = \tfrac{1}{6}n(n+1)(2n+1)$$

and
$$S_x^2 = \frac{\Sigma X^2}{n}-\left(\frac{\Sigma X}{n}\right)^2$$
$$= \frac{(n+1)(2n+1)}{6}-\frac{(n+1)^2}{4}.$$

Moreover, as $Y_1, Y_2, \ldots, Y_n$ represent the same numbers as $X_1, X_2, \ldots, X_n$ though not in the same order

$$S_X^2 = S_Y^2.$$

Now the rank differences $D_1, D_2, \ldots, D_n$ are given by

$$D_1 = (X_1 - Y_1), \quad D_2 = (X_2 - Y_2), \quad \ldots, \quad D_n = (X_n - Y_n)$$

and

$$\begin{aligned}\Sigma D^2 &= (X_1 - Y_1)^2 + (X_2 - Y_2)^2 + \ldots + (X_n - Y_n)^2 \\ &= (X_1^2 + X_2^2 + \ldots + X_n^2) - 2(X_1 Y_1 + X_2 Y_2 + \ldots + X_n Y_n) \\ &\qquad\qquad + (Y_1^2 + Y_2^2 + \ldots + Y_n^2) \\ &= \tfrac{1}{6}n(n+1)(2n+1) - 2\Sigma XY + \tfrac{1}{6}n(n+1)(2n+1).\end{aligned}$$

Thus $$\Sigma XY = \tfrac{1}{6}n(n+1)(2n+1) - \tfrac{1}{2}\Sigma D^2$$

and $$\begin{aligned}S_{XY} &= \frac{\Sigma XY}{n} - \left(\frac{\Sigma X}{n}\right)\left(\frac{\Sigma Y}{n}\right) \\ &= \tfrac{1}{6}(n+1)(2n+1) - \frac{1}{2n}\Sigma D^2 - \tfrac{1}{4}(n+1)^2.\end{aligned}$$

Hence the coefficient of rank correlation

$$\begin{aligned}\rho &= \frac{S_{XY}}{S_X S_Y} \\ &= \frac{\tfrac{1}{6}(n+1)(2n+1) - \tfrac{1}{4}(n+1)^2 - \dfrac{1}{2n}\Sigma D^2}{\tfrac{1}{6}(n+1)(2n+1) - \tfrac{1}{4}(n+1)^2} \\ &= 1 - \frac{\dfrac{1}{2n}\Sigma D^2}{\tfrac{1}{6}(n+1)(2n+1) - \tfrac{1}{4}(n+1)^2} \\ &= 1 - \frac{6\Sigma D^2}{n(n^2-1)}.\end{aligned}$$

Note that, where n is large, this formula approximates to

$$\rho = 1 - \frac{6\Sigma D^2}{n^3}$$

because n is negligible compared with n^3.

160. Method of ranking equal values of a variate. If the *cinema admissions (millions)* of §142 are arranged in descending order of magnitude they appear as follows:

11.0, 10.5, 10.0, 10.0, 9.9, 9.7, 9.4, 9.4, 9.3, 9.3, 9.0, 8.6.

The rank of 11.0 is, therefore, 1 and that of 10.5 is 2. The rank of the *two* values 10.0 is not taken as 3 or 4 but as 3.5, the mean of the ranks 3 and 4.

Similarly, the ranks of the two values 9.4 is taken as 7.5 and that of the two values 9.3 as 9.5. The twelve ranks are thus:

1, 2, 3.5, 3.5, 5, 6, 7.5, 7.5, 9.5, 9.5, 11, 12

and their sum is 78, the same as the sum of the ranks

1, 2, 3, 4, 5, 6, 7, 8, 9, 10, 11, 12

when the values of the variate are all different.

Table 13C shows the method of calculating the coefficient or rank correlation for the *cinema admissions*, x and *TV licences*, y, of § 142. It is a good example of inverse correlation. The value obtained, $\rho = -0.84$, is considerably higher (numerically) than the value $r_{xy} = -0.75$ obtained from the original (x, y) values of § 142. It must be clearly understood that table 12D is for minimum values of $|r_{xy}|$. Table 13D is a similar table for minimum values of $|\rho|$ prepared from M. G. Kendall's tables.

TABLE 13C

Calculation of the coefficient of rank correlation (*inverse*)

Cinema admissions (Millions) x	Cinema admissions Rank X	TV licences Per 1000 population y	TV licences Rank Y	Rank difference $X-Y$ D	D^2
11.0	1	12	12	-11	121
10.0	$3\frac{1}{2}$	18	11	$-7\frac{1}{2}$	$56\frac{1}{4}$
10.0	$3\frac{1}{2}$	24	10	$-6\frac{1}{2}$	$42\frac{1}{4}$
9.4	$7\frac{1}{2}$	37	9	$-1\frac{1}{2}$	$2\frac{1}{4}$
10.5	2	52	8	-6	36
9.7	6	64	7	-1	1
9.4	$7\frac{1}{2}$	69	6	$1\frac{1}{2}$	$2\frac{1}{4}$
9.3	$9\frac{1}{2}$	81	5	$4\frac{1}{2}$	$20\frac{1}{4}$
9.9	5	98	4	1	1
9.3	$9\frac{1}{2}$	101	3	$6\frac{1}{2}$	$42\frac{1}{4}$
9.0	11	106	2	9	81
8.6	12	119	1	11	121
			Total	0	$526\frac{1}{2}$

Since $n = 12$ and $\Sigma D^2 = 526\frac{1}{2}$

$$\rho = 1 - 6 \times 526\tfrac{1}{2}/12(12^2 - 1)$$

$$= -0.84.$$

161. Kendall's coefficient of rank correlation τ. Recently the theory and application of rank correlation methods have been considerably developed by M. G. Kendall in *Rank Correlation Methods*, published in 1948 and revised in 1955 and 1962. Kendall emphasizes that ranks have an import-

TABLE 13D

The $P = 5\%$ values of $|\rho|$

No. of pairs from which ρ is calculated n	Minimum value of $|\rho|$ for correlation to be likely
5	1
6	0.89
7	0.79
8	0.74
9	0.68
10	0.65
20	0.46

ance all their own. They fulfil functions which are more than mere approximations to the (x, y) pairs.* Kendall also introduces his coefficient of correlation denoted by the Greek letter τ which at first seems quite unrelated to r_{xy} or ρ, as it is calculated in a completely different way. It is interesting that H. E. Daniels in the *Journal of the Royal Statistical Society*, series B (1950) showed that

$$-1 \leqslant 3\tau - 2\rho \leqslant 1.$$

The full calculation of Kendall's τ for the rankings of vehicles and casualties is set out in table 13E. For convenience, the order of the years has been reversed so that the rankings R_x appear in ascending order and the rankings R_y are then slightly out of ascending order. Each (R_x, R_y) pair has been lettered in order *A, B, C, D, E, F, G, H, I, J*.

Consider every pair of letters that can be chosen from these letters then: *AB, AC, AD*, ..., *BC, BD, BE*, ..., *CD, CE, CF*, ... Altogether 45 such pairs are possible and in general, if there are n letters instead of 10, it can be proved that $\frac{1}{2}n(n-1)$ pairs are possible. The 45 pairs are set out according to an obvious system in table 13E and to each pair is attached a score, either $+1$ or -1. The way in which this score is obtained is illustrated by the following examples:

To obtain the score for AB

The ranks of R_x are 1, 2. These are in the *right* order and score $+1$.
The ranks of R_y are 1, 2. These are in the *right* order and score $+1$.
The product of the two *AB* scores is $(+1)(+1) = +1$.

To obtain the score for AG

The ranks of R_x are 1, 7. These are in the *right* order and score $+1$.
The ranks of R_y are 1, 9. These are in the *right* order and score $+1$.
The product of the two *AG* scores is $(+1)(+1) = +1$.

* Numerical values cannot be attached to *aesthetic appeal, flavour, degree of whiteness, surface polish*. Qualities such as these, however, can be used when obtaining ranks.

TABLE 13E

Calculation of Kendall's coefficient τ (direct)

R_x and R_y are reproduced from table 13B in ascending order for convenience.

	1954	1953	1952	1951	1950	1949	1948	1947	1946	1945	
	A	*B*	*C*	*D*	*E*	*F*	*G*	*H*	*I*	*J*	
R_x	1	2	3	4	5	6	7	8	9	10	Horizontal
R_y	1	2	4	3	5	6	9	7	8	10	totals
Pair		*AB*	*AC*	*AD*	*AE*	*AF*	*AG*	*AH*	*AI*	*AJ*	
Score		+1	+1	+1	+1	+1	+1	+1	+1	+1	+9
Pair			*BC*	*BD*	*BE*	*BF*	*BG*	*BH*	*BI*	*BJ*	
Score			+1	+1	+1	+1	+1	+1	+1	+1	+8
Pair				*CD*	*CE*	*CF*	*CG*	*CH*	*CI*	*CJ*	
Score				−1	+1	+1	+1	+1	+1	+1	+5
Pair					*DE*	*DF*	*DG*	*DH*	*DI*	*DJ*	
Score					+1	+1	+1	+1	+1	+1	+6
Pair						*EF*	*EG*	*EH*	*EI*	*EJ*	
Score						+1	+1	+1	+1	+1	+5
Pair							*FG*	*FH*	*FI*	*FJ*	
Score							+1	+1	+1	+1	+4
Pair								*GH*	*GI*	*GJ*	
Score								−1	−1	+1	−1
Pair									*HI*	*HJ*	
Score									+1	+1	+2
Pair										*IJ*	
Score										+1	+1
								Total score *S*			+39

Number of rankings, $n = 10.$

Maximum possible score, $\frac{1}{2}n(n-1) = 45.$

$$\text{Kendall's } \tau = \frac{S}{\frac{1}{2}n(n-1)}$$

$$= 0.87.$$

To obtain the score for CD

The ranks of R_x are 3, 4. These are in the *right* order and score +1.
The ranks of R_y are 4, 3. These are in *reverse* order and score −1.
The product of the two *CD* scores is $(+1)(-1) = -1$.

To obtain the score for GI

The ranks of R_x are 7, 9. These are in the *right* order and score +1.
The ranks of R_y are 9, 8. These are in *reverse* order and score −1.
The product of the two *GI* scores is $(+1)(-1) = -1$.

Finally, table 13E indicates that the total score, S, is 39 and Kendall defines τ by

$$\tau = \frac{\text{total score}}{\text{maximum possible score}}$$

$$= \frac{S}{\frac{1}{2}n(n-1)}.$$

TABLE 13F

Method of calculating Kendall's τ when the ranks R_x are in ascending order

R_y	Method of scoring	Score
1	Has 9 larger on its right	+9
2	Has 8 larger on its right	+8
4	1 smaller, 6 larger on its right	+5
3	Has 6 larger on its right	+6
5	Has 5 larger on its right	+5
6	Has 4 larger on its right	+4
9	2 smaller, 1 larger on its right	−1
7	Has 2 larger on its right	+2
8	Has 1 larger on its right	+1
	Total score, S	39
	Maximum possible score, $\frac{1}{2}n(n-1)$	45

$$\tau = \frac{39}{45}$$

$$= 0.87.$$

The complete procedure set out in table 13E can be shortened considerably. First of all it is evident that the score is +1 or −1 according as the two rankings *agree* or *disagree*. This eliminates the product idea of (+1)(+1), (+1)(−1), (−1)(−1). Secondly, because the R_x rankings are in strict ascending order it is only necessary to consider the R_y rankings and these can be assessed as shown in table 13F.

Table 13F is so much shorter than table 13E that it is evidently a good idea to put the ranks R_x in ascending order even though they are not in order when first obtained. Once the method of scoring is understood it need not be recorded as part of the calculation of τ.

Table 13G shows how tied ranks produce *zeros* in the scores. One example will be sufficient to illustrate the principle:

To obtain the score for BC

The ranks of R_x are $3\frac{1}{2}$, $3\frac{1}{2}$. These are *equal* and score 0.
The ranks of R_y are 11, 10. These are in *reverse* order and score −1.
The product of the two BC scores is $(0)(-1) = 0$.

TABLE 13G

Calculation of Kendall's τ with tied ranks (inverse)

R_x and R_y are reproduced from table 13C.

	A	B	C	D	E	F	G	H	I	J	K	L	Horizontal totals
R_x	1	$3\frac{1}{2}$	$3\frac{1}{2}$	$7\frac{1}{2}$	2	6	$7\frac{1}{2}$	$9\frac{1}{2}$	5	$9\frac{1}{2}$	11	12	
R_y	12	11	10	9	8	7	6	5	4	3	2	1	
Pair		*AB*	*AC*	*AD*	*AE*	*AF*	*AG*	*AH*	*AI*	*AJ*	*AK*	*AL*	
Score		−1	−1	−1	−1	−1	−1	−1	−1	−1	−1	−1	−11
Pair			*BC*	*BD*	*BE*	*BF*	*BG*	*BH*	*BI*	*BJ*	*BK*	*BL*	
Score			0	−1	+1	−1	−1	−1	−1	−1	−1	−1	−7
Pair				*CD*	*CE*	*CF*	*CG*	*CH*	*CI*	*CJ*	*CK*	*CL*	
Score				−1	+1	−1	−1	−1	−1	−1	−1	−1	−7
Pair					*DE*	*DF*	*DG*	*DH*	*DI*	*DJ*	*DK*	*DL*	
Score					+1	+1	0	−1	+1	−1	−1	−1	−1
Pair						*EF*	*EG*	*EH*	*EI*	*EJ*	*EK*	*EL*	
Score						−1	−1	−1	−1	−1	−1	−1	−7
Pair							*FG*	*FH*	*FI*	*FJ*	*FK*	*FL*	
Score							−1	−1	+1	−1	−1	−1	−4
Pair								*GH*	*GI*	*GJ*	*GK*	*GL*	
Score								−1	+1	−1	−1	−1	−3
Pair									*HI*	*HJ*	*HK*	*HL*	
Score									+1	0	−1	−1	−1
Pair										*IJ*	*IK*	*IL*	
Score										−1	−1	−1	−3
Pair											*JK*	*JL*	
Score											−1	−1	−2
Pair												*KL*	
Score												−1	−1
										Total score,		*S*	−47

Number of rankings, $n = 12.$

Maximum possible score, $\frac{1}{2}n(n-1) = 66.$

$$\text{Kendall's } \tau = \frac{S}{\frac{1}{2}n(n-1)}$$
$$= -0.71.$$

Table 13H, on p. 152, has been prepared from M.G. Kendall's tables. It will be found useful for making a decision about τ once it has been calculated.

162. Exercises.

1. Ten shades of the colour green when arranged in their true order from light to dark are numbered 1–10 respectively. An observer, when asked to arrange the shades from light to dark, produces the following rank

3, 1, 5, 2, 6, 4, 10, 9, 7, 8.

What is the value of Spearman's coefficient of rank correlation in this case? [London]

TABLE 13H

The $P = 5\%$ values of $|\tau|$

No. of pairs from which τ is calculated, n	Minimum value of $|\tau|$ for correlation to be likely
5	1
6	0.87
7	0.71
8	0.64
9	0.56
10	0.51
20	0.33

2. The following table gives the ages, in years and months, of ten boys together with the positions in which the boys were placed in a competition:

Position in competition	1	2	3	4	5	6	7	8	9	10
Age of competitor	14.3	14.0	14.9	14.5	12.11	13.5	13.8	13.11	13.1	11.10

Calculate the coefficient of correlation by ranks, and explain what inference you draw from it. [A.E.B.]

3. Ten shades of the colour red, when arranged in their true order from dark to light, were numbered 1 to 10 respectively. A boy, when asked to arrange the shades from dark to light, produced the following order:

1, 2, 5, 3, 4, 8, 6, 7, 10, 9.

Calculate, correct to three decimal places, the coefficient of correlation by ranks.

When two other boys carried out the same colour perception test, it was found that their coefficients of correlation were 0.192 and -0.972 respectively. Explain what inference might be drawn from each of the three coefficients. [A.E.B.]

4. Explain briefly what is meant by correlation.

A class of 15 students was examined in Mathematics and Physics. The following table gives the orders of merit of the students (arranged in alphabetical order) in both subjects. Calculate the coefficient of ranked correlation:

Mathematics	11	5	9	14	13	2	6	10	1	4	7	3	12	8	15
Physics	11	8	7	9	12	1	4	10	3	6	14	2	13	5	15

[London]

5. There are ten finalists in a competition for which there are two judges X and Y. The following table gives the order in which X and Y place the competitors.

Calculate the rank correlation coefficient and state any conclusions that you deduce from it:

Competitors	*A*	*B*	*C*	*D*	*E*	*F*	*G*	*H*	*K*	*L*
Ranking of *X*	6	10	2	7	1	4	5	9	8	3
Ranking of *Y*	2	7	8	10	6	1	9	5	3	4

[London]

6. In a drama competition ten plays were ranked by two adjudicators as follows:

Play	*A*	*B*	*C*	*D*	*E*	*F*	*G*	*H*	*J*	*K*
Rank given by *X*	5	2	6	8	1	7	4	9	3	10
Rank given by *Y*	1	7	6	10	4	5	3	8	2	9

Calculate the coefficient of ranked correlation. Is there any reason for saying that there is a significant agreement between the two adjudicators?

[London]

14

ANALYSIS OF VARIANCE CALCULATIONS

163. Significance of variance ratio. The procedure for applying the t-test in an investigation of the difference between the means of two small samples is given in chapter 9, §124. It is based on the assumption of a common parent variance, σ^2, of the two samples which is estimated by

$$\text{Est}\,(\sigma^2) = s^2 = \frac{(n_1-1)s_1^2+(n_2-1)s_2^2}{(n_1+n_2-2)},$$

where s_1^2 and s_2^2 are the sample variances. If, therefore, s_1 and s_2 differ so widely that the assumption of a common parent variance is not justified, the t-test is invalidated. Suppose that two samples of size n_1 and n_2 have variances s_1^2 and s_2^2 which have been calculated by using $\nu_1 = n_1-1$ and $\nu_2 = n_2-1$ degrees of freedom and that s_1^2 *is greater than* s_2^2.

The null hypothesis that the two samples are drawn from the same parent population of common variance s^2 *must be rejected at the* (*two-tailed*) 5 % *level if the variance ratio*

$$F = s_1^2/s_2^2$$

is greater than the (*one-tailed*) $P = 2\frac{1}{2}$ % *value of* F *to be found in table* A8, *page* 203.

Note that the F-tables A7 and A8 are arranged as *one-tailed* tests (i.e. for testing $s_1^2/s_2^2 > 1$ against the alternative hypothesis $s_1^2/s_2^2 = 1$). The test described above is two-tailed but we concentrate our attention on one-tail only when we select the larger of the two variances to calculate F as *larger variance/smaller variance.* The two examples of §§164 and 165 illustrate the method.

164. Example of a variance ratio which is significant. *The voltages of ten dry cells of type A are found to be*

1.50, 1.46, 1.52, 1.52, 1.47, 1.52, 1.49, 1.51, 1.49, 1.45

and those of eight cells of type B are

1.48, 1.46, 1.57, 1.54, 1.59, 1.46, 1.56, 1.58.

Is there reason to believe that the distributions of voltages in the two types of batteries are different and, if so, in what way? [A.E.B.]

To reduce the amount of arithmetic let us re-write the voltages with 1.50 as origin and 0.01 as unit.

Type *A*	x	0	−4	2	2	−3	2	−1	1	−1	−5
Type *B*	y	−2	−4	7	4	9	−4	6	8		

Then $\Sigma x = -7$, $\Sigma x^2 = 65$, $\bar{x} = -0.7$ and $s_x^2 = 6.678$ based on $\nu_x = 9$ degrees of freedom. Also $\Sigma y = 24$, $\Sigma y^2 = 282$, $\bar{y} = 3.0$ and $s_y^2 = 30$ based on $\nu_y = 7$ degrees of freedom. Because s_y^2 is greater than s_x^2 we take $s_1^2 = s_y^2 = 30$ with $\nu_1 = 7$ and $s_2^2 = s_x^2 = 6.678$ with $\nu_2 = 9$. Hence the variance ratio $F = s_1^2/s_2^2 = 4.49$. In table A8 we find that the $P = 2\frac{1}{2}\%$ value of F for $\nu_1 = 7$ and $\nu_2 = 9$ is 4.20. Thus the calculated value of F, 4.49, is greater than the $P = 2\frac{1}{2}\%$ value of F, 4.20. This means that the two-tail probability of obtaining such a large variance ratio from two samples drawn from a common variance population is less than 5 %. In other words, the variance ratio differs from unity at the 5 % level of significance and the null hypothesis that the two types of cell have a common variance must be rejected. The distributions of voltages in the two types of battery are, therefore, accepted as different. The voltages of type *B* batteries are more variable than those of type *A* batteries.

165. Example in which neither *F* nor *t* are significant. *In an experiment on the reaction times in seconds of two individuals A and B, found under identical conditions, the following results were obtained*:

A	0.41	0.38	0.37	0.42	0.35	0.38
B	0.32	0.36	0.38	0.33	0.38	

Examine the hypothesis that there is no difference between the distribution of reaction times for A and B. [A.E.B.]

Re-writing the times with 0.38 as origin and 0.01 as unit we get:

x	3	0	−1	4	−3	0
y	−6	−2	0	−5	0	

Hence $\Sigma x = 3$, $\Sigma x^2 = 35$, $\bar{x} = 0.5$, $s_x^2 = 6.7$ with $\nu_x = 5$, and $\Sigma y = -13$, $\Sigma y^2 = 65$, $\bar{y} = -2.6$, $s_y^2 = 7.8$ with $\nu_y = 4$. Taking $s_1^2 = s_y^2$ and $s_2^2 = s_x^2$ because $s_y^2 > s_x^2$ we get

$$F = s_1^2/s_2^2 = 1.16$$

which is less than the $P = 2\frac{1}{2}\%$ value of F, 7.39 for $\nu_1 = 4$ and $\nu_2 = 5$. Thus the variance ratio does not differ significantly from unity and the null hypothesis that the two samples have a common parent variance may be accepted.

We now proceed to apply the t-test to ascertain if the means of the two samples differ significantly. The procedure of chapter 9, § 124, gives $t = 1.91$ with $\nu = 9$. Thus the calculated value of t is less than the $P = 5\%$ value, 2.26, of table A5. The difference between the means is, therefore, not significant at the 5 % level and the hypothesis that there is no difference between the distribution of reaction times for A and B can be accepted with 95 % confidence.

166. Exercise.

The results of measurements on eight specimens of material A are respectively

92, 99, 96, 92, 98, 94, 99, 94 units

and the results of similar measurements on ten specimens of material B are respectively:

88, 91, 94, 87, 89, 91, 93, 93, 94, 96 units.

Do these figures indicate a convincing difference between the materials?
[A.E.B.]

167. Single factor analysis of variance. *In an examination in which the marks awarded formed approximately a normal distribution, there were ten candidates from each of three schools. The marks obtained were as follows:*

School A	51	24	45	30	13	64	15	9	21	91
School B	94	43	80	22	99	27	84	76	97	60
School C	65	49	1	6	67	2	55	18	49	32

Carry out an analysis of variance and examine the hypothesis that there is no essential difference between the schools. [A.E.B.]

This is an example of single factor analysis of variance because we are only interested in whether or not a variance exists between the means of the schools. If we can show that the existence of such a variance is *likely* then the hypothesis that 'there is no essential difference between the schools' must be rejected. Each school provides us with ten observations (or ten *replications*). Variances exist within these replications but, in this case, we are not interested in these. The ten replications are to be used to strengthen the evidence that differences between the schools either exist or do not exist. More confidence can be placed in a result derived from ten replications than, say, five. The method of analysis of variance allows for this fact. It will first be described with a detailed statement of the fundamental arithmetical processes. It will then be given in general terms with an explanation of the 'short cuts' that can be made in the arithmetic. In the description which follows note carefully the new terminology which is emphasised by the large capital letters.

I. First of all we ignore, for the time being, the meaning of the 3 *rows* and 10 *columns* and regard the 30 observations as one sample. Thus

(i) the grand AVERAGE $= (51+24+\ldots+49+32)/30$

$= 1389/30$

$= 46.3,$

(ii) the TOTAL-SUM-OF-SQUARES of deviations from the grand AVERAGE

$= (51-46.3)^2+(24-46.3)^2+\ldots+(32-46.3)^2$

$= 26954.3,$

(iii) the TOTAL-MEAN-SQUARE $= \dfrac{\text{total-sum-of-squares}}{\text{number of degrees of freedom}}$

$= 26954.3/29$

$= 929.5.$

II*a*. The second step is to eliminate the effect of the 10 replications in each of the 3 rows by replacing each by its *row-average*.

3 rows but only 2 D.F.	10 replications replaced by row average				
School *A*	36.3	36.3	36.3	...	(10 all alike)
School *B*	68.2	68.2	68.2	...	(10 all alike)
School *C*	34.4	34.4	34.4	...	(10 all alike)

This has the effect of concentrating the attention on the variation *between the schools* and eliminating the variation *within the schools.*

(i) The BETWEEN-THE-ROWS-SUM-OF-SQUARES of deviations of the 30 row-averages from the grand AVERAGE is then

$$10(36.3-46.3)^2+10(68.2-46.3)^2+10(34.4-46.3)^2 = 7212.2.$$

(ii) The BETWEEN-THE-ROWS-MEAN-SQUARE $= 7212.2/2$

$= 3606.1,$

because there are only 2 degrees of freedom (when 36.3 and 68.2 have been written down in the first column all the other 28 row-averages are determined by the total 1389). The beginner may be worried by the fact that 30 observations have only 2 degrees of freedom. Later in the calculation, however, when dealing with the components of variance we divide by the factor 10 to find σ_r^2, the between-the-rows variance.

II*b*. Finally, we deduct the BETWEEN-THE-ROWS-SUM from the TOTAL-SUM to obtain the WITHIN-THE-ROWS-SUM-OF-SQUARES and also the between-the-rows degrees of freedom from the total-degrees of freedom to obtain the within-the-rows degrees of freedom. Thus

(i) the WITHIN-THE-ROWS-SUM-OF-SQUARES $= 26954.3 - 7212.2$
$= 19742.1,$

(ii) the WITHIN-THE-ROWS-MEAN-SQUARE $= 19742.1/(29-2)$
$= 731.1.$

This is the best estimate of the variance that remains in the marks of the candidates after the effects of the different schools have been removed.

TABLE 14A

Single factor analysis
Components of variance

	Sum of squares	Degrees of freedom	Mean square	When the N.H. is accepted the mean square is an estimate of	When the N.H. is rejected the mean square is an estimate of
Between the rows ($r = 3$)	7212.2	2	3606.1	σ^2	$\sigma^2 + c\sigma_r^2$
Within the rows ($c = 10$)	19742.1	27	731.1	σ^2	σ^2
Total	26954.3	29	929.5	Best estimate of σ^2	—

Table 14A summarises the calculations made so far and shows how a conclusion is finally reached. If the null hypothesis that there is no essential difference between the schools is to be accepted then σ_r^2, the between-the-rows variance, must be zero. This implies that the mean squares 3606.1 and 731.1 are both equal because they are both estimates of σ^2. Hence their ratio will not differ significantly from unity. When the variance ratio test is applied we find

$$F = 3606.1/731.1$$
$$= 4.93$$

is greater than the $P = 5\%$ value 3.35 for $\nu_1 = 2$ and $\nu_2 = 27$.

The null hypothesis is, therefore, rejected.

Thus σ_r^2 is not zero, and, indeed, the equations

$$\sigma^2 + 10\sigma_r^2 = 3606.1,$$
$$\sigma^2 = 731.1,$$

can be solved to obtain $\sigma_r^2 = 287.5.$

If the three schools can be regarded as a random sample drawn from a large number of schools and the ten candidates in each case can be regarded as random samples, one from each school, then the results may be interpreted as follows:

(i) The distribution of a large number of candidates from *one school only* will be approximately normal with mean 46.3 and variance σ^2.

(ii) The distribution of a large number of candidates from a large number of schools will be approximately normal with mean 46.3 and variance $(\sigma^2+\sigma_r^2)$.

168. Better method of calculating the mean squares. The method of calculation described in the last paragraph illustrates the underlying meaning of 'between the rows' and 'within the rows' but it has serious drawbacks. Not only is it laborious; it is also erroneous. The AVERAGE and the row-averages may not be exact values. They may have to be stated to a given number of decimal places and the errors thus introduced, due to the squaring, will be cumulative. A more satisfactory method of calculation is given in table 14B. This shows the general case in which there are r rows and c columns (replications) the value x_{rc} representing the observation

TABLE 14B

Calculation of the mean squares
Single factor analysis

r rows but $(r-1)$ D.F.	c replications: 1	2	3	...	c	Row sum	Square of row sum	Row sum of squares
1	x_{11}	x_{12}	x_{13}	...	x_{1c}	R_1	R_1^2	S_1
2	x_{21}	x_{22}	x_{23}	...	x_{2c}	R_2	R_2^2	S_2
...	...	...	...	...	...	...	...	...
r	x_{r1}	x_{r2}	x_{r3}	...	x_{rc}	R_r	R_r^2	S_r
					Total	T	ΣR^2	S

Note that $$R_1 = (x_{11}+x_{12}+\dots+x_{1c})$$

and $$S_1 = (x_{11}^2+x_{12}^2+\dots+x_{1c}^2).$$

Also in the example being discussed $T = 1389$; $\Sigma R^2 = 715229$; $S = 91265$.

Cause	Sum of squares	Degrees of freedom
Between the rows	$(1/c)\,\Sigma R^2-(1/rc)\,T^2$	$(r-1)$
Within the rows	$S-(1/c)\,\Sigma R^2$	$(rc-r)$
Total	$S-(1/rc)\,T^2$	$(rc-1)$

Note that $(1/c)\Sigma R^2$ becomes $\Sigma(R^2/c)$ if c varies from row to row, i.e. if there are not the same number of replications in each row.

which is found in row r and column c. The reader will find it a useful exercise to go through the calculation following the pattern set out in table 14B and also to prove the formulae given for the sums of squares.

169. Assumptions made in the calculation. The single factor analysis which has been described is only valid if

(i) the row populations are normally distributed, and
(ii) the row variances are the same.

If these assumptions are in doubt they should be tested separately.

The *null hypothesis* tested by the analysis of variance calculation is that *the row means are the same.* By showing that the variance ratio differs significantly from unity we prove that the row means are significantly different. Some beginners are puzzled by this. It must be realised, however, that the squares involve differences of means and hence the appropriate sum of squares increases as the row means diverge. Thus the manipulation of the sums of squares leads to a test of the row means. Alternatively it will be understood that if the variance ratio differs significantly from unity then σ_r^2 is not zero. This implies that a variance does exist between the row means and so the row means are significantly different.

170. Exercises.

1. The following table shows the length of life in hours of 8 electric lamps drawn at random from the output of 6 different departments of a factory.

6 departments	Length of life in hours, 8 specimens							
	1	2	3	4	5	6	7	8
1	798	797	795	796	794	794	795	796
2	796	795	793	794	796	793	794	795
3	798	795	796	795	792	794	795	793
4	796	791	792	795	792	794	792	791
5	795	790	791	794	792	793	792	792
6	794	796	793	792	795	793	793	797

Is there any difference in the mean length of life of the lamps produced in the different departments?

2. In an experiment with rats to compare the carcinogenic effects of three nitrogen mustards, a score was used which may be assumed to be normally distributed. There were 5 rats in each group and the respective mean scores were 8, 12 and 4; the sum of the squares of all the scores was 1216. Carry out an analysis of variance to compare the three mustards, explaining in general (non-algebraic) terms how the manipulation of sums of squares can provide a test of differences of means. Explain any use you make of statistical tables. (Note that the data is equivalent to II*a* and I(ii).) [Cambridge specimen question]

171. Two-factor analysis of variance. The following example illustrates two-factor analysis of variance. In this case we are investigating the differences, if they exist, (i) between the row means and (ii) between the column means. Note that one observation only is given in each row-column cell.

Four workers are all tested for one hour on each of five machines, and the numbers of articles made are as follows:

	Machine				
Worker	1	2	3	4	5
A	21	24	26	19	25
B	29	30	27	24	35
C	17	22	23	16	22
D	25	24	32	25	34

Carry out an analysis of variance to ascertain whether there are significant differences (i) *between the workers,* (ii) *between the machines.*

[Institute of Statisticians]

First of all let us re-write the observations with 20 as origin and at the same time complete the row totals and column totals which will not be used until §172. As with single factor analysis the process will first be described by row averaging and column averaging. After this the formulae will be presented in §172 for the quick and generally more accurate calculation.

4 workers	5 machines ($c = 5$)							
($r = 4$)	1	2	3	4	5	Row sum	Square of row sum	Row sum squares of
A	1	4	6	−1	5	$R_1 = 15$	$R_1^2 = 225$	$s_1 = 79$
B	9	10	7	4	15	$R_2 = 45$	$R_2^2 = 2025$	$s_2 = 471$
C	−3	2	3	−4	2	$R_3 = 0$	$R_3^2 = 0$	$s_3 = 42$
D	5	4	12	5	14	$R_4 = 40$	$R_4^2 = 1600$	$s_4 = 406$
Column sum	$C_1 = 12$	$C_2 = 20$	$C_3 = 28$	$C_4 = 4$	$C_5 = 36$	$T = 100$	$\Sigma R^2 = 3850$	$S = 998$
Square of column sum	$C_1^2 = 144$	$C_2^2 = 400$	$C_3^2 = 784$	$C_4^2 = 16$	$C_5^2 = 1296$	$\Sigma C^2 = 2640$		

I. Ignoring, for the time being, the meaning of the rows and columns and regarding the 20 observations as one sample we get

(i) the grand AVERAGE $= (1+4+\ldots+5+14)/20$

$$= 5,$$

(ii) the TOTAL-SUM-OF-SQUARES of deviations from the grand AVERAGE is

$$(1-5)^2+(5-4)^2+\ldots+(5-5)^2+(14-5)^2 = 498,$$

(iii) the TOTAL-MEAN-SQUARE $= 498/19$

$$= 26.21.$$

II*a*. Eliminate the effect of the different machines by replacing each observation by its *row-average.*

4 Workers ($r = 4$) but only 3 D.F.	5 machines ($c = 5$)				
	1	2	3	4	5
A	3	3	3	3	3
B	9	9	9	9	9
C	0	0	0	0	0
D	8	8	8	8	8

(i) The BETWEEN-THE-ROWS-SUM-OF-SQUARES of deviations of the 20 row-averages from the grand AVERAGE is

$$5(3-5)^2+5(9-5)^2+5(0-5)^2+5(8-5)^2 = 270.$$

(ii) The BETWEEN-THE-ROWS-MEAN-SQUARE is $270/3 = 90$ because there are only 3 degrees of freedom.

II*b*. Eliminate the effect of the different workers by replacing each observation by its *column-average.*

4 workers ($r = 4$)	5 machines ($c = 5$)				
	1	2	3	4	5
A	3	5	7	1	9
B	3	5	7	1	9
C	3	5	7	1	9
D	3	5	7	1	9

5 columns but only 4 D.F.

(i) The BETWEEN-THE-COLUMNS-SUM-OF-SQUARES of deviations of the 20 column-averages from the grand AVERAGE is

$$4(3-5)^2+4(5-5)^2+4(7-5)^2+(1-5)^2+4(9-5)^2 = 160.$$

(ii) The BETWEEN-THE-COLUMNS-MEAN-SQUARE is $160/4 = 40$ because there are only 4 degrees of freedom.

II*c*. The RESIDUAL-SUM-OF-SQUARES is

(i) TOTAL-SUM − {BETWEEN-THE-ROWS-SUM + BETWEEN-THE-COLUMNS-SUM}

$$= 498-(270+160)$$
$$= 68.$$

(ii) The RESIDUAL-MEAN-SQUARE is

$$\frac{\text{RESIDUAL-SUM-OF-SQUARES}}{\text{TOTAL D.F.} - (\text{BETWEEN-THE-ROWS D.F.} + \text{BETWEEN-THE-COLUMNS D.F.})}$$
$$= 68/\{19-(3+4)\}$$
$$= 5.76.$$

The RESIDUAL-MEAN-SQUARE depends on the variation that still remains in the 20 observations after the effects of the different machines and different workers have been removed.

TABLE 14C

Two-factor analysis
Components of variance

	Sum of squares	Degrees of freedom	Mean square	When the N.H. is accepted the mean square is an estimate of	When the N.H. is rejected the mean square is an estimate of
Between the rows ($r = 4$)	270	3	90	σ^2	$\sigma^2 + c\sigma_r^2$
Between the columns ($c = 5$)	160	4	40	σ^2	$\sigma^2 + r\sigma_c^2$
Residual	68	12	5.67	σ^2	σ^2
Total	498	19	26.21	Best estimate of σ^2	—

The null hypothesis is that there is no difference (i) between workers, (ii) between machines. If (i) is true then $\sigma_r = 0$ and if (ii) is true then $\sigma_c = 0$ (Table 14C).

To ascertain whether or not $\sigma_r = 0$ we test the between-the-rows-mean-square as s_1^2 against the residual-mean-squares as s_2^2 and obtain

$$F = 90/5.67$$
$$= 15.9$$

which is greater than the $P = 5\%$ value 3.49 for $\nu_1 = 3$ and $\nu_2 = 12$.

(If the reader consults the *Cambridge Elementary Statistical Tables*, p. 11, the calculated value will be seen to be greater than the $P = 0.1\%$ value of F.) Hence part (i) of the null hypothesis is rejected; σ_r is not zero. To ascertain whether or not $\sigma_c = 0$ we test the between-the-columns-mean-square as s_1^2 against the residual-mean-square as s_2^2 and obtain

$$F = 40/5.67$$
$$= 7.06$$

which is greater than the $P = 5\%$ value 3.26 for $\nu_1 = 4$ and $\nu_2 = 12$.

(The *Cambridge Elementary Statistical Tables*, p. 10, indicate that the calculated value is greater than the $P = 1\%$ value of F.)

Hence part (ii) of the null hypothesis is rejected; σ_c is not zero. *The differences between the workers are significant at the* 0.1% *level. The differences between the machines are significant at the* 1% *level.* We may pursue the question further by solving the equations

$$\sigma^2 + 5\sigma_r^2 = 90,$$

$$\sigma^2 + 4\sigma_c^2 = 40,$$

$$\sigma^2 \qquad = 5.67,$$

thus obtaining $\sigma_r^2 = 16.87$, $\sigma_c^2 = 8.58$, $\sigma^2 = 5.67$.

In this case σ^2 is the variance we should expect in the number of articles per hour if *one worker* were to use *one machine* only for a large number of hours; $(\sigma^2 + \sigma_r^2)$ is the variance we should expect in the number of articles per hour if *one machine* were to be used in turn by a large number of workers for a large number of hours in each case; $(\sigma^2 + \sigma_c^2)$ is the variance we should expect in the number of articles per hour if *one worker* were to use a large number of machines, each for a large number of hours. Finally, $(\sigma^2 + \sigma_r^2 + \sigma_c^2)$ is the variance we should expect if *many* workers were to use *many* machines indiscriminately for *many* hours.

172. Better method of calculating the mean squares. If the original observations are given the general values

$$\begin{matrix} x_{11} & x_{12} & x_{13} & x_{14} & x_{15} \\ x_{21} & x_{22} & x_{23} & x_{24} & x_{25} \\ x_{31} & x_{32} & x_{33} & x_{34} & x_{35} \\ x_{41} & x_{42} & x_{43} & x_{44} & x_{45} \end{matrix}$$

the row-averages of II*a* being written

$$\begin{matrix} x_{1.} & x_{1.} & x_{1.} & x_{1.} & x_{1.} \\ x_{2.} & x_{2.} & x_{2.} & x_{2.} & x_{2.} \\ x_{3.} & x_{3.} & x_{3.} & x_{3.} & x_{3.} \\ x_{4.} & x_{4.} & x_{4.} & x_{4.} & x_{4.} \end{matrix}$$

and the column-averages of II*b* being written

$$\begin{matrix} x_{.1} & x_{.2} & x_{.3} & x_{.4} & x_{.5} \\ x_{.1} & x_{.2} & x_{.3} & x_{.4} & x_{.5} \\ x_{.1} & x_{.2} & x_{.3} & x_{.4} & x_{.5} \\ x_{.1} & x_{.2} & x_{.3} & x_{.4} & x_{.5} \end{matrix}$$

the reader will be able to prove the general formulae for mean squares given in table 14D.

(Note that $x_{r.}$ means the average of row *r* and that $x_{.c}$ means the average of column *c*.)

TABLE 14D

Calculation of the mean squares
Two factor analysis

r rows but $(r-1)$ D.F.	c columns but $(c-1)$ D.F. 1	2	3	...	c	Row sum $\sum_c x_{rc}$	Square of row sum	Row sum of squares $\sum_c x_{rc}^2$
1	x_{11}	x_{12}	x_{13}	...	x_{1c}	R_1	R_1^2	s_1
2	x_{21}	x_{22}	x_{23}	...	x_{2c}	R_2	R_2^2	s_2
⋮	⋮	⋮	⋮	⋮	⋮	⋮	⋮	⋮
r	x_{r1}	x_{r2}	x_{r3}	...	x_{rc}	R_r	R_r^2	s_r
Column sum $\sum_r x_{rc}$	C_1	C_2	C_3	...	C_c	T	ΣR^2	S
Square of column sum	C_1^2	C_2^2	C_3^2	...	C_c^2	Σc^2	—	—

Cause	Sum of squares	Degrees of freedom
Between the rows	$\frac{1}{c}\Sigma R^2 - \frac{1}{rc}T^2$	$(r-1)$
Between the columns	$\frac{1}{r}\Sigma C^2 - \frac{1}{rc}T^2$	$(c-1)$
Residual	$S - \frac{1}{c}\Sigma R^2 - \frac{1}{r}\Sigma C^2 + \frac{1}{rc}T^2$	$(rc-r-c+1)$
Total	$S-(1/rc)T^2$	$(rc-1)$

Early in § 171 it was established, for the example under discussion, that $T = 100$, $\Sigma R^2 = 5850$, $\Sigma C^2 = 2640$, $S = 998$, $r = 4$, $c = 5$. By substituting these values into the formulae given in table 14D the values of table 14C can be obtained. The reader will agree that this method of establishing table 14C is far more direct than that of setting out in full the row-averages and column-averages as first described in § 171. Note that

$$\frac{1}{c}\Sigma R^2 \quad \text{and} \quad \frac{1}{r}\Sigma C^2 \quad \text{become} \quad \Sigma\left(\frac{R^2}{c}\right) \quad \text{and} \quad \Sigma\left(\frac{C^2}{r}\right)$$

if the original table contains blank spaces, i.e. if some of the rc observations are missing.

173. When one of the between-mean-squares is not significant. Suppose that the final summary of our two factor analysis of variance were to be:

	Sum of squares	Degrees of freedom	Mean square
Between the rows $r = 4$	270	3	90
Between the columns $c = 5$	72	4	18
Residual	68	12	5.67
Total	410	19	21.57

We should find the between-the-rows-mean-square significant but the between-the-columns-mean square not significant. This implies that σ_c^2 is zero and 18 and 5.67 are both estimates of σ^2. A better estimate of σ^2 is obtained by combining the between-the-columns-sum-of-squares and the residual-sum-of-squares to give

$$\sigma^2 = (72+68)/(4+12)$$
$$= 8.75.$$

The value of σ_r^2 is then found by solving

$$\sigma^2 = 8.75 \quad \text{and} \quad \sigma^2+5\sigma_r^2 = 90.$$

174. When the residual-mean-square is significantly greater than either of the between-mean-squares. Suppose that the final summary of our two factor analyses of variance were to be:

	Sum of squares	Degrees of freedom	Mean square
Between the rows $r = 4$	270	3	90
Between the columns $c = 5$	12	4	3
Residual	216	12	18
Total	498	19	26.2

The residual-mean-square is significantly greater than the between-the-columns-mean-square. This is equivalent to saying that $\sigma^2 > \sigma^2+4\sigma_c^2$ which is absurd. One might suppose that it is the 5 % chance allowed for but it is safest to assume in this case that the readings are not independent. It does happen in experiments which are not properly designed that one reading can influence the next and so on. For example, if one reading is high it may cause the next reading to be either high or low. If the residual mean square is significantly greater than either of the between-mean-squares the fundamental design of the experiment should be carefully checked.

175. Exercises.

1. In a series of fatigue tests on metals, 5 different alloys were subjected to 5 different treatments. The following table is a summary of the results obtained.

Length of life in hours

5 different alloys	5 different treatments				
	A	*B*	*C*	*D*	*E*
a	1464	1459	1466	1468	1465
b	1462	1467	1467	1464	1476
c	1460	1462	1462	1468	1471
d	1469	1470	1468	1463	1474
e	1469	1472	1472	1472	1475

Use analysis of variance to test for differences between alloys and treatments.

2. An agricultural experiment was carried out by planting sixteen similar plots with four varieties of potatoes, and giving them four different manurial treatments, so that each variety and each treatment are used together in one plot.

The yields of potatoes in pounds per plot were as follows:

	Variety of potato			
Treatment	*A*	*B*	*C*	*D*
I	236	244	260	267
II	302	301	316	339
III	316	320	334	347
IV	296	309	321	316

Use analysis of variance to test for differences between varieties and between treatments. [A.E.B.]

3. Five types of flame photometer are to be compared by using them to measure the Na/K ratio of a solution of which ample quantities are available; five skilled technicians are available for the trial. Describe carefully an experimental design which will make it possible to compare photometers and technicians; explain in detail the procedure for setting up the experiment. Set out, in algebraic form, the analysis you would use. [Cambridge specimen question]

4. *The effect of climate on cloth*

5 different countries	4 different methods of testing			
	A	*B*	*C*	*D*
1	478	476	477	481
2	480	476	477	481
3	484	481	479	481
4	480	477	477	480
5	479	477	479	479

The above table shows the results of an experiment carried out to investigate the effect of climate on the tensile strength of cloth. A roll of cloth was divided into

5 parts and one part was dispatched to each of 5 laboratories in different parts of the world. Each laboratory divided its specimen into 4 sections and subjected them to tests *A*, *B*, *C*, *D*.

Use analysis of variance to test for differences between countries and between methods of testing.

5. What do you understand by degrees of freedom?

An experiment is carried out by 8 operatives on 6 machines, so that each operative uses each machine once only. Complete the following analysis of variance table and comment on the results.

Source of variation	Sums of squares	Degrees of freedom	Variance
Between machines	822		
Between operatives	1524		
Residual			
Total	4667		

[Institute of Statisticians]

6. Five doctors each test five treatments for a certain disease, and observe the number of days each patient takes to recover. The results are as follows (figures in days):

	Treatments				
Doctors	1	2	3	4	5
A	10	14	23	18	20
B	11	15	24	17	21
C	9	12	20	16	19
D	8	13	17	17	20
E	12	15	19	15	22

Using the method of analysis of variance, discuss the difference between (*a*) doctors and (*b*) treatments. [Institute of Statisticians]

15

MISCELLANEOUS EXERCISES

1. Part of a turbo-alternator test routine carried out at an Electric Power Station included observations of manometer readings at timed intervals. Two sets of readings were taken by two observers *A* and *B* using separate manometers. Both instruments were connected to the same orifice plate in the condensate line and observations were made alternately, at 15-sec intervals, by *A* and *B*. The following table summarises the two sets of readings made by the observers:

Manometer reading (mm of mercury)	Frequency	
	Observer *A*	Observer *B*
300–310	1	1
310–320	5	3
320–330	16	14
330–340	13	15
340–350	25	18
350–360	26	11
360–370	11	18
370–380	9	24
380–390	8	13
390–400	3	2
400–410	3	1
Total	120	120

Calculate the mean and the standard deviation of each set of readings.

2. Using the means and standard deviations obtained in Ex. 1 above, make out a table of expected frequencies (for the same class intervals as Ex. 1 above) based on the null hypothesis that each distribution is normal.

3. Using the observed and expected frequencies of Exs. 1 and 2 above, calculate values of χ^2 for each set of observations and hence show that the readings made by observer *A* did not differ significantly from normal, but those made by observer *B* did. Assuming that both observers were reliable what might be inferred about the manometers?

4. Obtain the 95 % confidence limits of the mean of the readings made by observer *A* in Ex. 1 above.

5. Estimate the level of significance of the difference between the means of the two sets of readings in Ex. 1 above.

6. A bridge player knows that his two opponents hold a total of 6 trump cards. If the cards are distributed at random, calculate the probability that neither of the opponents holds (i) more than 3 trump cards, (ii) more than 4 trump cards. [Cambridge]

7. In an examination the probability that candidate A will solve a given problem is $\frac{1}{4}$ and the probability that candidate B will solve it is $\frac{2}{3}$. What is the probability that the examiner will receive a correct solution from either A or B or from both assuming that they work independently? [Cambridge]

8. A bag contains 20 black balls and 15 white balls. (i) If 2 balls are withdrawn in succession, what is the probability that one is black and the other is white? (ii) If a sample of 5 balls is taken what is the probability that at least 4 are black? [Cambridge]

9. Two surveyors make repeated observations of the same angle. Their readings differ only in the figure for seconds of angle with the following results:

Surveyor	No. of observations	Mean reading (s)	Standard deviation (s)
A	6	38	4.33
B	8	34	3.21

Do these data provide evidence that there is a significant difference between the observations of A and B (i) at the 10 % level, and (ii) at the 5 % level of significance? [Cambridge]

10. The mean number of interruptions per hour to which a person is subjected during his normal working day is 4. Estimate the probabilities that in any particular hour he is interrupted (i) 4 times, (ii) not more than 4 times. [Cambridge]

11. Using the method of least squares, obtain the equation of the best straight line through the 3 following points:

x	0	1	2
y	2.3	6.5	10.6

[Cambridge]

12. An automatic machine produces bolts whose diameters are required to lie within the range 9.96 to 10.04 mm. A sample of 10 bolts is found to have a mean diameter of 9.98 mm. and a standard deviation of 0.02 mm. If the diameters are normally distributed, is there evidence, at the 5 % level, of an error in setting the machine?

If the machine is adjusted to produce bolts with a mean diameter of 10.00 mm, what proportion of bolts is likely to be rejected on full inspection? It may be assumed that the standard deviation of the diameters is not affected by the adjustment. [Cambridge]

13. It is found that on the average a certain electronic computer makes a mistake every 10 min. If the mistakes occur at random intervals, calculate the probability of no mistakes occurring in a calculation lasting (i) 5 min, (ii) 30 min. [Cambridge]

14. The relationship between the output voltage V in millivolts and the temperature difference θ in °C of a thermocouple is given approximately by the formula

$$V = a\left(\frac{\theta}{100}\right)+b\left(\frac{\theta}{100}\right)^2.$$

Use the method of least squares to derive equations which determine the best values of a and b from the following experimental results:

θ (°C)	100	200	300	400
V (millivolts)	0.95	1.40	1.12	0.20

[Cambridge]

15. The following contingency table was drawn up from the records of a firm which manufactures transformers:

	No. of transformers found satisfactory at first examination	No. of transformers found unsatisfactory at first examination
Department A	15	4
Department B	10	10

Do these figures provide conclusive evidence that the work done by department A is better than that done by department B?

16. Prove that, if f is the frequency of a measurement x, the mean of the measurements is M and x_0 is any number, then the standard deviation σ is given by

$$\sigma^2 = \frac{1}{N}\Sigma f(x-x_0)^2-(x_0-M)^2,$$

where $N = \Sigma f$. [London]

17. Show that the probability of exactly r successes in n trials is the coefficient of t^r in the expansion of $(1-p+pt)^n$, where p is the probability of success in any one trial.

On an expedition, a machine is taken which fails to start on the average once in c attempts owing to the breakage of a certain part. If s spare parts of this kind are carried, show that the probability that the last spare part will fail at the nth attempt at starting is equal to the coefficient of t^s in the expansion of

$$\frac{1}{c}\left(1+\frac{t-1}{c}\right)^{n-1}.$$

[London]

18. If the points with co-ordinates (x_1, y_1), (x_2, y_2), ..., (x_n, y_n) lie approximately on the straight line $y = ax+b$, show how to determine the values of the constants a and b which will make the sum

$$\sum_{r=1}^{n} (y_r-ax_r-b)^2$$

a minimum.

Find the equation of the straight line which satisfies this condition for the set of points with co-ordinates

$$(0, 13), \quad (3, 10), \quad (6, 8), \quad (9, 5), \quad (12, 2).$$

[London]

19. In an experiment which can succeed or fail, the probability of success is p and of failure is q, where $p+q = 1$. Show that the probability of r successful experiments in n attempts is given by the term in p^r in the binomial expansion of $(q+p)^n$.

In the manufacture of screws by a certain process it was found that 5 % of the screws were rejected because they failed to satisfy tolerance requirements. What was the probability that a sample of 12 screws contained (i) exactly 2, (ii) not more than 2 rejects? [London]

20. Use the method of least squares to find the values of a and b which nearly satisfy the four equations

$$2a+b = 2.76, \qquad 4a+3b = 3.90,$$
$$3a-b = 8.62, \qquad a+2b = -1.38.$$

[London]

21. A competition consists of filling in a form which contains N spaces and each space has to be filled up in one of n ways. There is a unique correct solution. If the spaces are filled up in a random manner, prove that the probability of there being r mistakes is the coefficient of x^r in the expansion of

$$n^{-N}\{1+x(n-1)\}^N.$$

If $N = 12$ and $n = 3$ prove that the probability of there being *not more* than 2 mistakes is 289.3^{-12}. [London]

22. If σ is the standard deviation and μ the mean of the data in a given statistical table, and if σ_1^2 is the mean of the squares of the deviations from an assumed mean a, show that

$$\sigma^2 = \sigma_1^2-(\mu-a)^2.$$

The annual salaries of a group of employees are given in the following table, in which £S is the salary and N is the number receiving each amount:

S	450	500	550	600	650	700	750	800
N	3	5	8	7	9	7	4	7

Calculate the mean salary and standard deviation. [London]

23. If the points with co-ordinates (x_i, y_i), $(i = 1, 2, \ldots, n)$ lie approximately on a straight line $y = mx+c$, find the values of m and c in terms of Σx, Σy, Σx^2 and Σxy if the sum $\sum_{i=1}^{n} (y_i-mx_i-c)^2$ is a minimum.

Hence find the values of m and c for the case where (x_i, y_i) are given by (1, 14), (2, 11), (3, 8), (5, 4), (6, 0) and deduce the best value of y corresponding to $x = 4$. [London]

24. The yield of a chemical process was measured at 3 temperatures, each with 2 concentrations of a particular reactant, as recorded below:

Temperature (t °C)	40	40	50	50	60	60
Concentration (x)	0.2	0.4	0.2	0.4	0.2	0.4
Yield (y)	38	42	41	46	46	49

Use the method of least squares to find the best values of the coefficients a, b, c in the equation

$$y = a+bt+cx,$$

and from your equation estimate the yield at 70 °C with concentration 0.5. [London]

HINT: To obtain the three normal equations for a, b and c differentiate $\Sigma(y-a-bt-cx)^2$ (i) with respect to a, (ii) with respect to b, (iii) with respect to c and equate each differential coefficient to zero.

25. The mean of 50 readings of a variable was 7.43, and their standard deviation was 0.28. The following 10 additional readings become available: 6.80, 7.81, 7.58, 7.70, 8.05, 6.98, 7.78, 7.85, 7.21, 7.40.

If these are included with the original 50 readings find (i) the mean, (ii) the standard deviation of the whole set of 60 readings. [London]

26. Calculate the mean and variance of the binomial distribution in which the distribution of the relative frequencies (the total frequency is unity) of 0, 1, 2, ..., n successes in n events consists of the terms in the expansion of $(q+p)^n$, where $q+p = 1$, q being the chance of failure and p of success in each event.

Two ounces of seeds of yellow wallflowers are thoroughly mixed before sowing with 8 oz of seeds of red wallflowers and eventually the plants are bedded out in rows of 20. Estimate the mean and variance of the number of yellow flowers in each row. [Northern]

27. Calculate the mean value of $10 \cos \theta$ over the range $0 \leqslant \theta \leqslant \pi/2$, giving the result in terms of π.

The distance between adjacent printed lines on a sheet of writing paper is 1 cm. The paper is laid flat on a horizontal table and a straight needle 10 cm long is thrown at random 40 times on to the paper. The following table shows the frequency distribution of the number of intersections of the printed lines made by the needle as it lies on the table after each throw:

No. of intersections	0	1	2	3	4	5	6	7	8	9	10
Frequency	3	4	1	3	1	1	4	4	6	6	7

Calculate the mean number of intersections.

Combine the mean value obtained in the first part of this question with the mean obtained in the second to estimate the value of π, explaining by means of a diagram the principle involved. [Northern]

28. A variable x is distributed at random between the values 0 and 1 so that the equation of the frequency curve is

$$y = Ax^2(1-x)^3,$$

where A is a constant. Find the value of A such that the area under the frequency curve $\int_0^1 y\,dx$ is unity.

Using this value of A determine

(i) the mean, $\bar{x} = \int_0^1 xy\,dx$,

(ii) the variance, $\sigma^2 = \int_0^1 x^2y\,dx-\bar{x}^2$.

NOTE: For a *continuous distribution* such as this $\int y\,dx$, $\int xy\,dx$ and $\int x^2y\,dx$ correspond to Σf, Σfx and Σfx^2 in a discrete frequency distribution.

29. A variable x is distributed at random between the values 0 and 4 so that the equation of the frequency curve is

$$y = Ax^3(4-x)^2,$$

where A is a constant. Find the value of A such that the area under the frequency curve is unity. Determine the mean and standard deviation of the distribution. [Northern]

30. A variate x can assume values only between 0 and 1, and the equation of its frequency curve is

$$y = Ae^{-2x} \quad (0 < x < 1),$$

where A is a constant such that the area under the curve is unity. Determine the value of A to three decimal places. Calculate the mean and variance of the distribution. [Northern]

31. A variate x can assume values only between 0 and a, and the equation of its frequency curve is

$$y = A(a-x)^2 \quad (0 \leqslant x \leqslant a),$$

where A and a are constants such that the area under the curve and the mean of the distribution are both unity. Determine the numerical values of A and a and find the 10th and 90th percentiles of the distribution. [Northern]

32. If n is a positive integer show that $\int_0^\infty x^n e^{-x}\,dx = n!$.

A variate has for its frequency distribution curve the graph of $y = xe^{-x/a}$ for $x > 0$, where $a > 0$. Find the total frequency, the mean and the standard deviation of the variate. [London]

33. The length x of the side of a square is rectangularly distributed between 1 and 2. Show that the area y of the square is distributed between 1 and 4 with a probability distribution $p(y)\,dy = \frac{1}{2}y^{-\frac{1}{2}}\,dy$.

Sketch the frequency curve, and calculate the mean and variance of the area of the square.

HINT: For the rectangular distribution, $p(x)\,dx = A\,dx$ (A constant).

But $\int_1^2 p(x)\,dx = 1$ and thus $A = 1$. Now $x = y^{\frac{1}{2}}$ and hence

$$\int_1^2 dx = \int_1^4 \tfrac{1}{2}y^{-\frac{1}{2}}\,dy.$$

34. The length x of the edge of a cube is rectangularly distributed between 5 and 10. Show that the volume y of the cube is distributed between 125 and 1000 with a probability distribution $p(y)\,dy = \frac{1}{15}y^{-\frac{2}{3}}\,dy$.

Sketch the frequency curve, and calculate the mean and variance of the volume of the cube. [Northern]

35. The radius x of a circle is rectangularly distributed between 1 and 2. Show that the area y of the circle is distributed between π and 4π with probability distribution

$$p(y)\,dy = \tfrac{1}{2}\pi^{-\frac{1}{2}}y^{-\frac{1}{2}}\,dy.$$

Sketch the probability curve and calculate the mean and the variance of the area of the circle. [Northern]

36. (i) A variate x can take the values $1, 2, 3, \ldots, n$ with equal probability. Calculate the standard deviation of x.

(ii) The standard deviation of two samples, each of n observations, are s_1 and s_2, their respective means being m_1 and m_2. Show that the standard deviation s of the combined sample of $2n$ observations is given by

$$s^2 = \tfrac{1}{2}(s_1^2+s_2^2)+\tfrac{1}{4}(m_1-m_2)^2.$$ [Northern]

37. (i) A variate x can take the values $1, 2, 3, \ldots, n$ with probabilities proportional to $1, 2, 3, \ldots, n$ respectively. Calculate the mean value of x.

(ii) Two samples, consisting of n_1 and n_2 observations, have means m_1 and m_2 and standard deviations s_1 and s_2 respectively. Show that the standard deviation s of the combined sample of (n_1+n_2) observations is given by

$$s^2 = \frac{n_1 s_1^2+n_2 s_2^2}{n_1+n_2}+\frac{n_1 n_2(m_1-m_2)^2}{(n_1+n_2)^2}.$$ [Northern]

38. When A and B play chess, the chance of either winning a game is $\frac{1}{4}$ and the chance of the game being drawn is $\frac{1}{2}$. Find the chance of A winning at least three games out of five.

Also, if A and B play a match to be decided as soon as either has won two games, find the chance of the match being finished in ten games or less.

[Northern]

39. If $\phi(x) = \dfrac{1}{\sqrt{(2\pi)}}e^{-\frac{1}{2}x^2}$, show that $x^n\phi(x) \to 0$ when $x \to \infty$.

Evaluate
$$\int_{-\infty}^{\infty} |x|\;\phi(x)\,dx.$$

Also, assuming that

$$\int_{-\infty}^{\infty} \phi(x)\,dx = 1, \quad \text{evaluate} \quad \int_{-\infty}^{\infty} x^2\phi(x)\,dx.$$

Hence obtain, for the distribution $\phi(x)$, the ratio of the mean deviation to the standard deviation. [Northern]

40. The two equal sides of an isosceles triangle are each of unit length and the angle θ between them is rectangularly distributed between 0 and $\pi/6$. Show that the area y of the triangle is distributed between 0 and $\frac{1}{4}$ with probability distribution

$$p(y)\,dy = \frac{12}{\pi}(1-4y^2)^{-\frac{1}{2}}\,dy.$$

Sketch the probability curve and calculate the mean and variance of the area of the triangle. [Northern]

41. The journey by air from one city centre A to another city centre B is divided into five stages and the means and standard deviations of the time taken for each stage are shown in the following table:

The five stages of the journey	Mean time (min)	Standard deviation (min)
Motor-coach journey from city centre *A* to air terminal *A*	35	2
Wait at air terminal *A*	12	2
Air journey from air terminal *A* to air terminal *B*	65	4
Wait at air terminal *B*	8	1
Motor-coach journey from air terminal *B* to city centre *B*	25	2

Assuming that the times taken for each stage of the journey are independently normally distributed, calculate the mean and standard deviation of the time taken for the whole journey.

Estimate the probability of the whole journey taking (i) less than $2\frac{1}{4}$ h, (ii) more than $2\frac{3}{4}$ h. [Northern]

42. In a certain school the age of each of 500 boys and his percentage mark in a general knowledge test were recorded. The relationship between the number pairs thus obtained is shown in the following table:

Percentage mark	Age in years (x)									
(y)	10	11	12	13	14	15	16	17	18	19
80–89	—	—	—	—	—	—	—	2	—	—
70–79	—	—	—	—	3	2	6	3	4	1
60–69	—	—	—	10	15	26	19	14	2	—
50–59	—	2	7	32	43	23	7	2	0	1
40–49	—	2	28	50	31	15	2	1	—	—
30–39	—	10	32	38	6	1	—	—	—	—
20–29	—	11	28	4	—	—	—	—	—	—
10–19	3	7	7	—	—	—	—	—	—	—

(i) Calculate the coefficient of correlation r_{xy} for the age in years x and the percentage mark y.

(ii) Obtain, by the method of least squares, the equation

$$y = mx + c$$

of the line of regression of y on x. [Northern]

43. A hundred samples, each of 12 articles, are drawn at random from a large bulk of articles, all of which should be alike, but 10% of which are in fact defective. Estimate the number of samples in which you would expect the number of defective articles to be (i) 0, (ii) 1, (iii) 2, (iv) 3 or more. [Northern]

44. A man who was trying out various makes of razor blade kept a record of the number of shaves x he got from each blade before he had to discard it. For 50 blades of one make he found that x was approximately normally distributed with mean 7.2 and standard deviation 2.0. Draw the frequency curve for this distribution, taking a scale of 1 in. to two shaves per blade and making the area under the curve 5 sq in.

For 50 blades of a different make he obtained the following distribution:

x	5	6	7	8	9	10	11	12	13	14
Frequency	1	2	4	6	9	9	8	6	3	2

On the same diagram as before draw a histogram of area 5 sq in. to represent the new distribution. Describe in words any inference that can be made from the two distributions. [Northern]

45. Calculate the number of different selections of three letters that can be made from the 10 letters of the word MANCHESTER.

Find, in addition, the probability that any one selection of three letters will contain (i) just one E, (ii) at least one E.

46. The following table is taken from a recent report on the serving of school dinners in grammar schools of Great Britain:

Time taken for school dinner in minutes	20	25	30	35	40	Total
No. of girls' schools	5	19	125	65	49	263
No. of co-educational schools	10	35	121	56	44	266

Calculate the mean and standard deviation of the times taken (i) in girls' schools, (ii) in co-educational schools.

Determine whether or not the difference between the means is significant.

47. In determining the nature and seriousness of an illness a common hospital procedure is to measure the *Erythrocyte Sedimentation Rate* (*E.S.R.*), using a solution of sodium citrate and fresh venous blood from the patient. The citrate prevents the blood from clotting and is called the *anticoagulant*. For other haematological tests sequestrin is often used as the anticoagulant. The question arises: Is it satisfactory to determine the *E.S.R.* by citrating blood which has already been collected in sequestrin? In order to try to answer this question the *E.S.R.*'s of 20 patients, all seriously ill, were determined (i) by using sodium citrate alone, and (ii) by using sodium citrate mixed with sequestrin. The following table gives the results:

	Patient	1	2	3	4	5	6	7	8	9	10
E.S.R. (i) (sod. cit. alone)	x	79	38	26	77	37	125	68	22	117	32
E.S.R. (ii) (sod. cit. + sequestrin)	y	83	44	45	81	41	125	69	24	121	31

	Patient	11	12	13	14	15	16	17	18	19	20
E.S.R. (i) (sod. cit. alone)	x	82	20	32	41	99	81	105	145	145	6
E.S.R. (ii) (sod. cit. + sequestrin)	y	80	20	38	47	78	83	114	143	143	6

Calculate the mean and the standard deviation of the differences between x and y and determine whether or not the mean of the differences differs significantly from zero.

Suppose it is decided that the readings of patients no. 3 and no. 15 are erroneous and that they should be ignored, determine whether or not the mean of the differences of the other 18 pairs of readings differs significantly from zero.

[Northern]

48. Given that the mean and the standard deviation of the n values

$$x_1, \quad x_2, \quad x_3, \quad \ldots, \quad x_n$$

are $\bar{x}$ and s respectively, obtain the mean and the standard deviation of the n values

$$(ax_1+b), \quad (ax_2+b), \quad (ax_3+b), \quad \ldots, \quad (ax_n+b).$$

The marks gained by a group of candidates in a certain subject in an examination had a mean of 42 and a standard deviation of 7. Obtain a formula for converting the marks so as to make the mean 50 and the standard deviation 10.

If a scholarship is to be awarded on the marks obtained in three subjects, explain why conversions such as the above applied to the three separate sheets of marks would help in making the award. [Northern]

49. A variate x can assume values only between 0 and 5 and the equation of its frequency curve is

$$y = A \sin \tfrac{1}{5}\pi x \quad (0 \leqslant x \leqslant 5),$$

where A is a constant such that the area under the curve is unity. Determine the value of A and obtain the median and quartiles of the distribution.

Show also that the variance of the distribution is

$$50\left\{\frac{1}{8}-\frac{1}{\pi^2}\right\}.$$

[Northern]

50. A box contains 36 batteries of which 4 are defective. If a random sample of 5 batteries is drawn from the box, calculate, correct to three places of decimals, the probability of the number of defective batteries in the sample being (i) 0, (ii) 1, (iii) more than 1. [Northern]

ANSWERS TO EXERCISES

§6. **1.** 981.2; 0.74. **2.** 1.040 kg, 0.0296 kg; 1.020 kg, 0.0157 kg.

3. 0.745 cm, 0.029 cm; 0.723 cm, 0.012 cm. **4.** 32 min, 4.43 min.

5. 10, $\sqrt{2}$; $\frac{1}{2}\sqrt{3}$. **6.** 10.2.

7. 49.5; 7.4. **8.** $\dfrac{n_1 M_1 + n_2 M_2}{n_1 + n_2}$; $\left\{\dfrac{n_1\sigma_1^2 + n_2\sigma_2^2}{n_1+n_2}\right\}^{\frac{1}{2}}$.

9. 8.83; 83.3; 34.5. **10.** 2.1 (late); 3.3. $5\frac{1}{2}$ (late); 1 (early).

11. 50, 20, 21, 43, 61, 74. **12.** $y = (\frac{5}{4})x - 35$; $\bar{y} = (\frac{5}{4})\bar{x} - 35$; $s' = \frac{5}{4}s$.

13. $5\frac{5}{6}$; 6.8. **14.** 9. **15.** 6; 29.2.

§9. **1.** 0.81 cm/s². **2.** 4.95 min. **3.** 0.0311 kg; 0.0165 kg.

4. 0.031 cm; 0.013 cm.

§18. **2.** 45.0; 16.6 (16.3 Sheppard). **3.** 42.74; 4.05.

4. 15.20; 9.46 (9.35 Sheppard). **6.** 228 kg/cm²; 21 kg/cm².

7. 371.3 m; 20.3 m (20.1 Sheppard). **8.** 163.9 cm; 1.6 cm.

9. 21.6; 1.8. **10.** 10.67; 2.29. To fall towards 10.5.

12. 100.5; 10.13; 13.4. **13.** 57.1.

14. 3.87; 1.92. **15.** 0; 1. *D*, *A*, *C*, *B*.

§29. **1.** (i) 0.01; (ii) 0.61; (iii) 9.94; (iv) 39.44; (v) 39.44; (vi) 9.94; (vii) 0.61; (viii) 0.01.

2. 0.03, 0.79, 7.26, 26.38, 38.11, 21.95, 5.01, 0.47.

3. 0.13, 8.99, 53.94, 34.67, 2.22, 0.01.

4. 9, 63, 156, 131, 37, 4.

§32. **1.** (i) 1.97; (ii) 5.29. **2.** (i) 9.12; (ii) 2.28.

3. 60 mm, 0.2432 mm. (i) 10.9 over + 10.9 under; (ii) 1.99 over + 1.99 under.

4. 108 ohms, 92 ohms. **6.** £52.53.

§43. **1.** $\frac{11}{24}$. **2.** (i) $\frac{23}{100}$; (ii) $\frac{379}{1800}$.

3. 0.46. **4.** $\frac{1}{120}$. **5.** (i) $\frac{1}{1296}$; (ii) $\frac{19}{1296}$; (iii) $\frac{523}{648}$.

6. $\frac{216}{671}$, $\frac{180}{671}$, $\frac{150}{671}$, $\frac{125}{671}$. Total = 1.

7. (i) $\frac{1}{506}$; (ii) $\frac{6}{506}$; (iii) $\frac{320}{506}$.

8. (i) $m/(m+n)$; (ii) $m(m-1)/(m+n)(m+n-1)$;
(iii) $n/(m+n-1)$; (iv) $m/(m+n)$.

9. (*a*), (*c*), (*d*) all fair; (*b*) unfair.

10. (*a*) $\frac{1}{12}$; (*b*) $\frac{5}{648}$; (*c*) $\frac{31}{216}$.

§46. **1.** $\frac{48}{89}$. **2.** $\frac{2}{5}$. **3.** $\frac{1}{3}$. **4.** $\frac{5}{9}$.
5. $\frac{1}{14}$. **6.** $\frac{1}{3}$. **7.** $\frac{1}{6}$.

§54. **1.** (i) 3136; (ii) 2296; $\frac{3}{8}$. **2.** $\frac{9}{14}$.

3. (i) $2^9.3^{16}.17.19/50^{20} = 0.07$;
(ii) $3^{10}(2^{12}.11.13.17.19.+.2^9.3^6.17.19+3^{10})/5^{20} = 0.19$.

4. $\frac{8}{35}$; $\frac{1}{2}$;

1	$\frac{1}{2}$	0	$-\frac{1}{2}$	-1
$\frac{1}{70}$	$\frac{16}{70}$	$\frac{36}{70}$	$\frac{16}{70}$	$\frac{1}{70}$

.

5. $\frac{1}{4}$; $\frac{21}{100}$. **6.** (i) $\frac{1}{285}$; (ii) $\frac{1}{57}$; (iii) $\frac{32}{57}$.

§57. **1.** $\frac{125}{216}$, $\frac{75}{216}$, $\frac{15}{216}$, $\frac{1}{216}$; 125, 75, 15, 1.

2. $\frac{3125}{7776}$, $\frac{3125}{7776}$, $\frac{1250}{7776}$, $\frac{250}{7776}$, $\frac{25}{7776}$, $\frac{1}{7776}$. 40.2, 40.2, 16.1, 3.2, 0.3, 0.0.

§59. $\frac{1}{16}$, $\frac{4}{16}$, $\frac{6}{16}$, $\frac{4}{16}$, $\frac{1}{16}$; 2.5, 10, 15, 10, 2.5.

§61. **1.** 59.0, 32.8, 7.3, 0.9. **2.** 60.3, 30.9, 7.5, 1.3.

§64. **1.** (i) 0.15; (ii) 0.96. **2.** 4, 11, 14, 11, 6, 3, 1.

3. (i) 0.0055; (ii) 0.2903. **4.** $\frac{864}{3125}$; $\frac{513}{625}$.

5. (i) $\frac{1}{3375}$; (ii) 0.339. **6.** $\frac{280}{2187} = 0.128$.

§69. **1.** 0.24, 0.048. **2.** $18\frac{3}{4}$ %.

§71. (i) 0.132, 0.266, 0.268, 0.178, 0.089, 0.067. (ii) 0.135, 0.271, 0.271, 0.180, 0.090, 0.053.

§76. **1.** 60.7 %, 9.0 %. **2.** 0.9048, 0.0905, 0.0047.

3. 0.3679, 0.3679, 0.1840, 0.0613, 0.0189.

4. 0.6065, 0.3033, 0.0902. **5.** (i) 0.0498; (ii) 0.3526.

6. 0.0069. **7.** 0.6703. **8.** $m = 1.87$, $s^2 = 1.76$.

9. $m = 0.128$, $s = 0.366$; 20941, 2680, 172, 7, 0, 0.

10. $m = 1.182$, $s = 1.167$, 397 or 398.

11. 22.3 %, 19.1 %, 16.7 %, 18.6 %.

§84. **1.** $m = 1.076, s = 0.25$; E's are 14, 53, 123, 155, 106, 40, 9. $\nu = 4, \chi^2 = 20.4$, distribution probably not normal.

2. $m = 908.6$, $s = 53.55$; E's are 23, 121, 305, 341, 169, 41. $\nu = 3$, $\chi^2 = 1.33$, distribution probably normal.

3. $m = 1714$, $s = 66.1$; E's are 6, 36, 124, 249, 291, 197, 77, 20. $\nu = 5$, $\chi^2 = 2.0$, distribution probably normal.

4. $m = 118.4, s = 1.71$; E's are 5, 78, 333, 415, 152, 17. $\nu = 3, \chi^2 = 24.6$, distribution probably not normal.

§86. **1.** $p = 0.1875, n = 6$; E's are 11.5, 16, 12.5; $\nu = 1, \chi^2 = 0.04$, distribution probably binomial.

2. $p = \frac{1}{12}$, $n = 12$; E's are 17, 19, 14; $\nu = 1$, $\chi^2 = 0.13$, distribution probably binomial.

3. E's are 14, 22, 16, 8; $\nu = 3, \chi^2 = 0.54$, probably binomial.

4. E's are 11, 32, 40, 26, 11; $\nu = 4, \chi^2 = 5.35$, probably binomial.

5. E's are 26, 130, 260, 260, 130, 26; $\nu = 5, \chi^2 = 3.36$, probably binomial.

§89. **1.** E's are 14, 27, 25, 15, 11; $\nu = 3, \chi^2 = 2.46$, probably Poisson.

2. E's are 20941, 2680, 172, 7, 0, 0; $\nu = 2, \chi^2 = 23$, probably not Poisson.

3. E's are 336, 398, 235, 93, 27, 7; $\nu = 4, \chi^2 = 18$, probably not Poisson.

4. E's are 13, 15, 12; $\nu = 1, \chi^2 = 0.14$, probably Poisson.

5. E's are 18, 18, 14; $\nu = 1, \chi^2 = 0.13$, probably Poisson.

6. E's are 16, 21, 14, 9; $\nu = 2, \chi^2 = 0.16$, probably Poisson.

7. E's are 16, 32, 32, 22, 18; $\nu = 3, \chi^2 = 7.6$, probably Poisson.

8. E's are 68, 171, 213, 179, 111, 90; $\nu = 4, \chi^2 \doteqdot 140$, probably not Poisson.

§97. **1.** (i) $\nu = 1, \chi^2 = 3$, differences not significant, no definite improvement.
(ii) $\nu = 1, \chi^2 = 6.6$, differences significant, definite improvement.

2. No. $\chi^2 = 3.2$, differences not significant.

3. No. $\chi^2 = 3.5$, differences not significant.

§99. **1.** $\chi^2 = 0.04$, probably no association exists.

2. $\chi^2 = 13.1$, association probably exists.

§100. **1.** $\chi^2 \doteqdot 119$. There was probably association.

2. $\chi^2 = 9.9$. Association probably does exist. Different mortality rates are probably associated with different types of anaesthetic.

3. No. $\chi^2 = 2.5$. Association probably does not exist.

4. Yes. $\chi^2 \doteqdot 15$. Association probably does exist.

5. Yes. $\chi^2 = 56$. Association most probably exists.

§102. 1. $\nu = 4$, $\chi^2 = 4.9$, probably no association.

2. $\nu = 6$, $\chi^2 = 48.3$, very probably association.

3. E's are 98, 146, 140, 103, 186, 280, 266, 200, 54, 82, 78, 56; $\nu = 6$, $\chi^2 = 162$, association very probable indeed.

4. E's are all 1000. $\nu = 9$ (totals of O's and E's agree). $\chi^2 = 9.32 < P =$ 10 % value 14.68. Differences not significant. Hypothesis accepted. $\chi^2 = 1.98 < P = 99$ % value 2.09. Remarkable agreement between E's and O's.

5. E's are 256, 1024, 1536, 1024, 256. $\nu = 4$ (totals of O's and E's agree). $\chi^2 = 2.05 < P = 10$ % value 7.78. Differences not significant. Hypothesis accepted.

6. E's (based on N.H. that all days are the same) are all 258. $\nu = 5$, $\chi^2 = 20 > P = 0.5$ % value 16.75. Differences very significant. N.H. rejected. Strong evidence that more books are generally borrowed on one weekday than on another.

7. E's based on N.H. that the proportions of the various kinds of workers are the same for all three firms: 24, 48, 48, 36, 72, 72, 40, 80, 80. $\nu = (h-1)(k-1) = 4$, $\chi^2 = 2.07 < P = 10$ % value 7.78. Differences not significant. No significant difference between the proportions of the various kinds of workers.

8. E's based on the N.H. that the proportions of passes and failures are the same for all three schools: 47.7, 25.3, 81.8, 43.2, 25.5, 13.5, $\nu = (h-1)(k-1) = 2$. $\chi^2 = 0.376 < P = 10$ % value 4.61. Differences not significant. N.H. accepted. Actual numbers (rather than percentages) give the true state of affairs.

9. (i) O's: 27, 34, 6, 11, 11, 11. E's: 25.46, 30.15, 11.39, 12.54, 14.85, 5.61. $\nu = (h-1)(k-1) = 2$. $\chi^2 = 9.5 > P = 1$ % value 9.21. Differences significant. N.H. (that there is no difference between the schools) is rejected. There is strong evidence that some schools are better than others at Pure Mathematics.

(ii) The N.H. that there is no association between passing in Pure and passing in Applied implies that the E's must be calculated by assuming that the results for Pure are independent of the results for Applied. O's are: 67, 55, 33, 45. E's are: 61, 61, 39, 39. $\nu = 1$ (2×2 table). $\chi^2 = 3.03$ without using Yate's correction or $\chi^2 = 2.54$ using Yate's correction. Differences not significant. N.H. is accepted.

10. E's are: 18.75, 37.5, 18.75, 6.25, 12.5, 6.25, 25, 50, 25. $\nu = (h-1)(k-1) = 4$. $\chi^2 = 48 > P = 0.1$ % value 18.47. Very significant. N.H. rejected. Very marked association between hair colour and eye colour.

11. E's are: 108.3, 121.9, 37.9, 56.9, 78.4, 88.1, 27.4, 41.1, 13.3, 15.0, 4.7, 7.0. $\nu = (h-1)(k-1) = 6$. $\chi^2 = 36.5 > P = 0.1$ % value 22.46. Very significant. Association between number employed and condition of factory is extremely likely.

12. (*a*) E's are: 2248, 392, 1069, 5848, 1018, 2781 (based on N.H. of no association between school and university). $\nu = 2$. $\chi^2 = 2167$, extremely large. N.H. rejected. It is extremely likely that the type of school determines the type of university in the case of men.

(*b*) E's are: 337, 78, 118, 2932, 683, 1023. $\nu = 2$. $\chi^2 = 157$, very large. It is very likely that the type of school determines the type of university but the likelihood is less marked in the case of women than in the case of men.

§108. 1. (*a*) 1.040 ± 0.039 K; (*b*) 0.85 approx.

2. (*a*) 0.745 ± 0.046 cm; (*b*) (i) $\frac{1}{4}$ approx.; (ii) $\frac{1}{8}$ approx.

3. 507.3 ± 0.96; 1460.

§111. 1. Yes; $t = 2.36$. 2. No; $t = 1.77$. 3. No; $t = 1.67$.

§113. 1. $\nu = 9$, $t = 1.16$, difference not significant.

2. $\nu = 11$, $t = 2.79$, difference is significant. The machine is set high.

§118. 1. 13.70, 0.5093. 2. 3 min 49 s, 2.015 s; 0.0236.

3. 0.06, 0.0384; 79.2 %.

§123. 1. $t = 2.24$; significant at 5 % level but not at 1 % level.

2. $t = 2.12$; difference is significant; Also the standard deviations prove that the thickness is nowhere less than 0.2 in.

§126. 1. $m_1 = 11.88$, $s_1^2 = 53.5$; $m_2 = 13.08$, $s_2^2 = 66.5$. S.E. = 2.5, $t = 0.48$, difference not significant.

2. By the mean of the differences, $\nu = 9$, $t = 2.2$ slightly less than $P = 5\%$ value. Improvement very doubtful. By difference of means $\nu = 18$, $t = 0.82$. Improvement not at all significant.

3. A: $t = 3.58 > P = 1\%$ value; significant (reads high). B: $t = 3.36 > P = 1\%$ value; significant (reads low).

4. $\nu = 10$, $t = 2.39 > P = 5\%$ value. Significant. Experiment would be sounder using method of matched pairs of chickens.

5. $\nu = 13$, $t = 4.29 > P = 0.1\%$ value. Highly significant. The control group had higher ability in the first place. Improvement probably would be more difficult for them.

6. $\nu = 5$, $t = 2.325 < P = 5\%$ valuc. Difference not significant.

7. $\nu = 10$, $t = 0.64$. No significant difference. A: $\nu = 5$, $t = 1.82$. Not significantly below 1000. B: $\nu = 5$, $t = 1.9$. Not significantly below 1000.

8. $t = 3.79$; out of control. 9. Yes. 110.4, 127.8; 6.

10. $t_4 = 2.795$; yes. 11.1 tons. 11. $t = 2.6$; yes. Type II.

12. $t_9 = 4.12$. Very effective.

13. $t_{11} = 3.5$, significant at 1% level. Take paired readings.

14. $t_7 = 3.05$, significant at 2% level.

15. $t_9 = 5.45$, significant at 0.1% level.

16. $t_9 = 4.76$, significant at 0.2% level. **17.** $t_{16} = 4.12$; 2.57

18. $t_7 = 5.446$. **19.** $t_{13} = 1.914$. Use estimate based on all 15.

20. Not significant at 5% level; significant at 5% level.

21. 20.4; 0.69. $t_9 = 1.83$; not.

22. 0.0041; 0.0042; 1.499 ± 0.003.

§133. (i) 10.86; (ii) 5.9; (iii) 10.86 ± 2.22; (iv) 10.86 ± 3.50. Under control after 1 week; out of control after 4 weeks.

§135. 97.5% zone 10.7; 99.9% zone 13.8. Two elevens indicate that control is doubtful after 1 week. 14, 12, 13, 15 indicate production completely out of control after 4 weeks.

§137. Allowable width 95.3 to 104.7 (i.e. 5.3 to 14.7 above 90). (i) Satisfactory after 1 week; (ii) not satisfactory after 4 weeks.

§139. $m = 1.5$; $c = 4$; as there are 4 *or more* 3 times out of 40 production is under control with a process average of $7\frac{1}{2}$% defective. But the last 6 samples indicate that it is probably going out of control.

§151. **1.** $x = -28.5y + 2225$. $y = -0.03x + 75$.

2. $x = 1.54y + 1.53$. **3.** $y = 33.3x - 7.0$.
12.3%; 10.75 across, 11.4 along.

4. $r = 2.135$, $R = 0.055$. **5.** $y = 54.9 + 0.5x$.

6. $y = 0.216x - 9.25$, 0 is farthest above the line.

8. $y = 109.5 + 2.716x - 0.183x^2$. $dy/dx = 0$ when $x = 7.5$.

§156. **1.** -0.76. **2.** -0.93. **3.** 0.86. **4.** 0.82.

5. 0.83. **6.** 0.91; $y = 0.9x + 2$. (i) 24.5; (ii) 42.5; (iii) 60.5.

7. 0.66. $y = 0.39x + 57$; £252.

§162. **1.** 0.78. **2.** 0.73. **3.** 0.915. **4.** 0.8; 0.64.

5. 0.02; 0.07. **6.** 0.62; 0.42.

§166. $F = 1.05$, not significant. $t = 2.63$, significant.

§170. **1.** $F = 4.46$, significant. **2.** $F = 10$, significant.

§175. **1.** Alloys, $F = 4.8$, significant. Treatments, $F = 3.8$, significant.

2. Treatments, $F = 134$, significant. Varieties, $F = 21$, significant.

4. Countries, $F = 4.33$, significant. Methods, $F = 7.85$, significant.

5. Machines, $F = 2.48$, not significant. Operatives, $F = 3.27$, significant.

6. Doctors, $F = 2.99$, not significant. Treatments, $F = 47$, significant.

ANSWERS TO EXERCISES

Miscellaneous Exercises

1. 351, 356, 21.5, 21.8 (mm Hg). 21.3, 21.5 using Sheppard's correction.

2. A: 3, 5, 11, 17, 21, 22, 18, 12, 7, 3, 1.
B: 2, 4, 8, 14, 20, 22, 20, 15, 9, 4, 2.

3. A: $\chi^2 = 9.47$, $\nu = 5$; not significant at 5% level.
B: $\chi^2 = 12.58$, $\nu = 4$; significant at 2.5% level.

4. 351 ± 42 (mm Hg).

5. $t = 1.7$; significant at 9% level.

6. (i) $\binom{6}{3}\binom{20}{10}\Big/\binom{26}{13}$, (ii) $\left\{\binom{6}{2}\binom{20}{11} + \binom{6}{3}\binom{20}{10} + \binom{6}{4}\binom{20}{9}\right\}\Big/\binom{26}{13}$.

7. $\frac{3}{4}$.

8. (i) $\frac{60}{119}$; (iii) $\frac{741}{2728}$.

9. $s = 3.718$, $t = 1.99$, $\nu = 12$. Significant at 10% level, but not at 5% level.

10. (i) $4^4e^{-4}/4! = 0.1952$; (ii) 0.6283.

11. $y = 4.15x+2.32$.

12. $t = \sqrt{10}$. Significant 4.56%.

13. (i) $e^{-0.5} = 0.6065$; (ii) $e^{-3} = 0.0498$.

14. $a = 1.32$, $b = -0.316$.

15. $\chi^2 = 1.71$, $\nu = 1$; not significant.

18. $a = -0.9$; $b = 13$.

19. (i) 0.0987; (ii) 0.9797.

20. $a = 2.3$, $b = -1.79$.

22. £636; £103.4.

23. $m = -2.66$, $c = 16.45$; 5.8.

24. $a = 18.92$, $b = 0.38$, $c = 20$; 55.

25. (i) 7.44; (ii) 0.30.

26. 4, 3.2.

27. $20/\pi$; 6.2, $\pi = 3.23$.

28. $A = 60$, $\bar{x} = \frac{3}{7}$, $\sigma^2 = \frac{3}{98}$.

29. $A = \frac{15}{1024}$, $\bar{x} = 2\frac{2}{7}$, $\sigma = 2\sqrt{6/7}$.

30. $A = 2.314$, $\bar{x} = 0.343$, $\sigma = 0.263$.

31. $A = \frac{3}{64}$, $a = 4$; 0.138, 2.143.

32. a^2; $\bar{x} = 2a$, $\sigma = a\sqrt{2}$.

33. $\bar{y} = 2\frac{1}{3}$, $\sigma^2 = \frac{34}{45}$.

34. $\bar{y} = 468\frac{3}{4}$, $\sigma^2 = 63755$.

35. $\bar{y} = 7\pi/3$, $\sigma^2 = 34\pi^2/45$.

36. $\sqrt{\{(n^2-1)/12\}}$.

37. $(2n+1)/3$.

38. $\frac{53}{512}$, $\frac{1981}{2048}$.

39. $\sqrt{\frac{2}{\pi}} = 0.8$.

40. $\frac{3}{2\pi}(2-\sqrt{3})$;

$$\frac{1}{16\pi^2}(2\pi^2-3\sqrt{3}\pi-252+144\sqrt{3}) = 0.0005.$$

41. 2 h 25 min, 5.385 min; (i) 0.0316, 0.0001.

42. (i) 0.75; (ii) $y = 6.63x-42.9$.

43. (i) 28; (ii) 38; (iii) 23; (iv) 11.

45. 92; (i) $\frac{7}{15}$; (ii) $\frac{8}{15}$.

46. (i) 32.55, 4.682; (ii) 31.67, 5.105. $t = 2.065$; significant.

47. 2, 7.1; $t = 1.26$, not significant. 1.9, 3.2; $t = 2.66$, significant.

48. $a\bar{x}+b$, as. $y = 10(x/7-1)$.

49. $\frac{1}{10}\pi$; 2.5, $1\frac{2}{3}$, $3\frac{1}{3}$.

50. (i) $\frac{899}{1682}$; (ii) $\frac{4595}{11781}$; (iii) $\frac{331}{3927}$.
(i) 0.534; (ii) 0.382; (iii) 0.084.

GLOSSARY

This not only summarises the ideas of chapters 1 to 14 but also extends them.

α error. See 'Type I error'.

Analysis of variance. The calculation of the components of two-factor analysis of variance is summarised below:

r rows	c columns: 1	2	...	c	Row sum $\sum_c x_{rc}$	Square of row sum	Row sum of squares $\sum_c x_{rc}^2$
1	x_{11}	x_{12}	...	x_{1c}	R_1	R_1^2	s_1
2	x_{21}	x_{22}	...	x_{2c}	R_2	R_2^2	s_2
⋮	⋮	⋮	⋮	⋮	⋮	⋮	⋮
r	x_{r1}	x_{r2}	...	x_{rc}	R_r	R_r^2	s_r
Column sum $\sum_r x_{rc}$	C_1	C_2	...	C_c	T	ΣR^2	S
Square of column sum	C_1^2	C_2^2	...	C_c^2	ΣC^2	—	—

Cause	Sum of squares	Degrees of freedom	Mean square is an estimate of
Between the rows	$\frac{1}{c}\Sigma R^2 - \frac{1}{rc}T^2$	$(r-1)$	$\sigma^2 + c\sigma_r^2$
Between the columns	$\frac{1}{r}\Sigma C^2 - \frac{1}{rc}T^2$	$(c-1)$	$\sigma^2 + r\sigma_c^2$
Residual	$S - \frac{1}{c}\Sigma R^2 - \frac{1}{r}\Sigma C^2 + \frac{1}{rc}T^2$	$(rc-r-c+1)$	σ^2
Total	$S-(1/rc)T^2$	$(rc-1)$	—

Note that $\frac{1}{c}\,\Sigma R^2$ becomes $\Sigma\,\frac{R^2}{c}$ if c varies from row to row and similarly $\frac{1}{r}\,\Sigma C^2$ becomes $\Sigma\frac{C^2}{r}$ if r varies from column to column.

Bayes' theorem. Let $A_1, A_2, \ldots, A_n$ be a mutually exclusive and exhaustive set of outcomes of a random process, and B be a chance event such that $P(B) \neq 0$ then

$$P(A_r|B) = \frac{P(B|A_r)P(A_r)}{\sum_{r=1}^{n} P(B|A_r)P(A_r)}.$$

β error. See 'Type II error'.

Bernoulli's theorem. The theorem of which the binomial distribution is a corollary. If the probability of an event occurring at a single trial is p, the probability of exactly r occurrences of the event in n independent trials is

$$\binom{n}{r} p^r(1-p)^{n-r}.$$

Binomial distribution. Suppose that the probability of an event occurring at a single trial is p and the probability of it not occurring is q then $p+q = 1$ and the probabilities

$$P(0),\quad P(1),\quad P(2),\quad \ldots,\quad P(r),\quad \ldots,\quad P(n)$$

of $0, 1, 2, \ldots, r, \ldots, n$ occurrences of the event in n independent trials are given by the terms of the *binomial expansion*

$$(q+p)^n = q^n + \binom{n}{1} q^{n-1}p + \binom{n}{2} q^{n-2}p^2 + \ldots + \binom{n}{r} q^{n-r}p^r + \ldots + p^n.$$

The mean value of r, $\mu = np$ and the variance of r, $\sigma^2 = npq$. As n increases the binomial distribution tends to the normal. The minimum size of n for which the normal distribution is a close approximation to the binomial distribution depends on the value of p. The normal approximation to the binomial is quite good provided np and $n(1-p)$ are both greater than 5.

Central limit theorem. If random samples of size n and mean $\bar{x}$ are drawn from a population with mean μ and variance σ^2 the distribution of $(\bar{x}-\mu)/(\sigma/\sqrt{n})$ approaches *normal* (0, 1) as n tends to infinity.

(Note that the approximation is quite good even when n is relatively small. For some parent populations $n \geqslant 15$ is necessary but for others $n \geqslant 5$ may be satisfactory.)

Chebyshev's theorem. Given a probability distribution with mean μ and variance σ^2 the probability of getting a value which differs from μ by more than $k\sigma$ is less than $1/k^2$. Alternatively, the probability of getting a value which differs from μ by less than $k\sigma$ is more than $1-1/k^2$.

Chi-squared (Pearson's).

$$\chi^2 = \Sigma\left[\frac{(O-E)^2}{E}\right],$$

where O is the observed frequency of a particular class and E is the corresponding expected frequency.

Coefficient of correlation.

$$r_{xy} = \frac{\text{Covariance}}{\text{Product of standard deviations}} = \frac{s_{xy}}{s_x s_y}.$$

r_{xy} differs significantly from zero (5 % level) if

$$t = r_{xy}\sqrt{\left(\frac{n-2}{1-r_{xy}^2}\right)} \quad \text{with} \quad \nu = n-2$$

is greater than the P (= 5 %) value of t in table A 5. For a given value of n,

$$r_{xy} = t/\sqrt{(t^2+n-2)} \quad \text{with} \quad \nu = n-2$$

gives the minimum value of $|r_{xy}|$ for correlation to be probable (see table 12 D, page 142).

Coefficient of rank correlation (Spearman's).

$$\rho = 1 - \frac{6\Sigma D^2}{n(n^2-1)}.$$

Coefficient of regression. The gradient of the regression line.
The coefficient of regression of y on x is s_{xy}/s_x^2.
The coefficient of regression of x on y is s_{xy}/s_y^2.

Combinations (or selections) $\binom{n}{r}$ **or** ${}^nC_r$. The number of combinations of n unlike things taken r at a time is

$$\binom{n}{r} = \frac{n(n-1)(n-2)\ldots(n-r+1)}{1.2.3\ldots r} = \frac{n!}{r!(n-r)!}.$$

Confidence limits of the mean. The 95 % confidence limits of the mean are $m \pm (ts/\sqrt{n})$, where t is the $P = 5\,\%$ value of the t-distribution for $\nu = n-1$ (sample of size n). Similarly the 99.8 % confidence limits are obtained from the $P = 0.2\,\%$ value. The probability that the true mean lies outside the 95 % confidence limits is 5 % or $\frac{1}{20}$.

Covariance of a bivariate distribution.
For n separate (x, y) pairs

$$s_{xy} = \frac{1}{n}\Sigma(x-\bar{x})(y-\bar{y})$$

$$= \frac{\Sigma xy}{n} - \left(\frac{\Sigma x}{n}\right)\left(\frac{\Sigma y}{n}\right).$$

For a grouped distribution

$$s_{xy} = \frac{\Sigma fxy}{\Sigma f} - \left(\frac{\Sigma f_x x}{\Sigma f}\right)\left(\frac{\Sigma f_y y}{\Sigma f}\right).$$

Degrees of freedom. In calculating the standard deviation of n observations the sum of the deviations from the mean is zero. Hence, when $(n-1)$ deviations have been written down, the nth deviation is determined and we say that for n observations there are $(n-1)$ degrees of freedom. See also table A3, page 198.

Exponential distribution. Suppose that incidents occur independently and at random intervals, the probability of an incident occurring in any interval δx being $\lambda\delta x$. The probability distribution of the intervals between consecutive incidents, $p(x) = \lambda e^{-\lambda x}$, for which the mean interval between incidents is $1/\lambda$, is called the exponential distribution.

***F*-test.** Suppose that two samples of size n_1 and n_2 have variances s_1^2 and s_2^2 which have been calculated by using $\nu_1 = n_1 - 1$ and $\nu_2 = n_2 - 1$ degrees of freedom and that $s_1^2 > s_2^2$. If the variance ratio $F = s_1^2/s_2^2$ is greater than the $P = 2\frac{1}{2}\,\%$ value of F for ν_1 and ν_2 given in table A8 then the null hypothesis that the two samples are drawn from populations with the same variance must be rejected at the 5 % level. The *Cambridge Elementary Statistical Tables* also give $P = 1\,\%$ and $P = 0.1\,\%$ values of F.

Interval estimate. A range of values which indicates the reliability of the point estimate of a parameter θ. If we can find two functions L and U of the sample values such that $P\{L < \theta < U\} = 0.95$ then the interval estimate (L, U) is a 0.95, or 95 %, confidence interval for θ.

Least squares line of best fit. If $y = a+bx$ is the equation, the values of a and b are given by the *normal* equations

$$\Sigma y = na+b\Sigma x,$$

$$\Sigma xy = a\Sigma x+b\Sigma x^2,$$

where n is the number of (x, y) pairs.

An alternative form of the equation is

$$(y-\bar{y}) = \frac{s_{xy}}{s_x^2}(x-\bar{x}),$$

or

$$(y-\bar{y}) = \frac{r_{xy}s_y}{s_x}(x-\bar{x}).$$

Predictions of y based on the above equation can only be regarded as expectations. If α and β are the respective population values corresponding to a and b, the two-tail percentage points of the t-distribution, with $\nu = (n-2)$ degrees of freedom, can be used to obtain interval estimates of β, α and y. First calculate s_e, the standard error of estimate either by

$$s_e^2 = \frac{\sum_{r=1}^{n}(y_r-a-bx_r)^2}{n-2},$$

or by

$$s_e^2 = \frac{\Sigma y^2-a\Sigma y-b\Sigma xy}{n-2}.$$

Then

(i) $t = \dfrac{|b-\beta|}{s_e}\sqrt{\left(\dfrac{n(\Sigma x^2)-(\Sigma x)^2}{n}\right)}$ gives the significance of $|b-\beta|$. Note that b is said to be significant if it differs significantly from zero.

(ii) $b \pm \dfrac{ts_e}{\sqrt{\left(\dfrac{n(\Sigma x^2)-(\Sigma x)^2}{n}\right)}}$ gives the confidence limits for β.

(iii) $t = \dfrac{|a-\alpha|}{s_e\sqrt{\left(\dfrac{1}{n}+\dfrac{n\bar{x}^2}{n(\Sigma x^2)-(\Sigma x)^2}\right)}}$ gives the significance of $|a-\alpha|$.

Note that if a does not differ significantly from zero the line passes through the origin.

(iv) $a \pm ts_e\sqrt{\left(\dfrac{1}{n}+\dfrac{n\bar{x}^2}{n(\Sigma x^2)-(\Sigma x)^2}\right)}$ gives confidence limits for α.

(v) $(a+bx_0) \pm ts_e\sqrt{\left(\dfrac{1}{n}+\dfrac{n(x_0-\bar{x})^2}{n(\Sigma x^2)-(\Sigma x)^2}\right)}$ gives confidence limits for the *mean value of* y estimated from x_0.

(vi) $(a+bx_0) \pm ts_e\sqrt{\left(1+\frac{1}{n}+\frac{n(x_0-\bar{x})^2}{n(\Sigma x^2)-(\Sigma x)^2}\right)}$ gives confidence limits for an *individual value of y* estimated from x_0.

Level of significance. See 'Significance'.

Lower quartile. The lower quartile divides the area under the probability curve (not necessarily 'normal') in the ratio 1:3. The lower quartile is the 25th percentile.

Mean. The arithmetic mean, or more simply the mean, of the n values $x_1, x_2, \ldots, x_n$ is

$$\bar{x} = \frac{1}{n}(x_1+x_2+\ldots+x_n)$$

$$= \frac{1}{n}\Sigma x.$$

If the n values have respective frequencies $f_1, f_2, \ldots, f_n$

$$\bar{x} = \frac{f_1x_1+f_2x_2+\ldots+f_nx_n}{f_1+f_2+\ldots+f_n}$$

$$= \Sigma fx/\Sigma f.$$

For a *continuous probability curve* $y = f(x)$ $(a \leqslant x \leqslant b)$,

$$\int_a^b y\,dx = 1 \quad \text{and} \quad \mu = \int_a^b xy\,dx.$$

Mean deviation. The mean deviation of the n values $x_1, x_2, \ldots, x_n$ is

$$\frac{1}{n}\{|x_1-\bar{x}|+|x_2-\bar{x}|+\ldots+|x_n-\bar{x}|\} = \frac{1}{n}\Sigma|x-\bar{x}|.$$

If the n values have respective frequencies $f_1, f_2, \ldots, f_n$ the mean deviation

$$= \frac{f_1|x_1-\bar{x}|+f_2|x_2-\bar{x}|+\ldots+f_n|x_n-\bar{x}|}{f_1+f_2+\ldots+f_n}$$

$$= \Sigma f|x-\bar{x}|/\Sigma f.$$

For a *continuous probability curve* $y = f(x)$ $(a \leqslant x \leqslant b)$, mean deviation

$$= \int_a^b |x-\bar{x}|\,y\,dx.$$

Median. The median bisects the area under the probability curve (not necessarily 'normal'). The median is the 50th percentile.

Normal equations. See 'Least squares line of best fit'.

Normal frequency curve. The equation of the normal frequency curve in its most general form is

$$y = \frac{1}{\sigma\sqrt{(2\pi)}}e^{-\frac{1}{2}(x-\mu)^2/\sigma^2}.$$

Here, μ is the mean and σ the standard deviation and the distribution is known as *normal* (μ, σ). Table A1, page 195, gives values of the normal (0, 1) distribution.

Null hypothesis. The null hypothesis is the *assumption* which is made when applying a significance test.

Parameters. Constants appearing in the specification of probability distributions, e.g. p (binomial), a (Poisson), μ, σ^2 (normal), λ (exponential).

Percentiles. The values which divide the area under the probability curve (not necessarily 'normal') into a hundred equal parts. The 25th, 50th and 75th percentiles are known as the lower quartile, median and upper quartile respectively.

Point estimate. A single numerical value obtained from a random sample which locates a parameter θ such as the mean, standard deviation or correlation coefficient as a single point on the real number scale. A point estimate does not indicate the reliability or precision of the method of estimation. For this we need an interval estimate. The ideal point estimate $\hat{\theta}$ will be *unbiased* and of minimum variance. Also as the sample size increases the estimate $\hat{\theta}$ should converge towards θ. In this case $\hat{\theta}$ is called *consistent*.

Poisson distribution. The Poisson distribution is the form assumed by the binomial distribution when p is small and n is large, the mean number of occurences np being a finite constant a. In this case the probabilities $P(0)$, $P(1), P(2), ..., P(r)$ of $0, 1, 2, ..., r$ occurrences of the event are

$$e^{-a}, \quad ae^{-a}, \quad \frac{a^2}{2!}e^{-a}, \quad ..., \quad \frac{a^r}{r!}e^{-a}$$

respectively.

Power of a significance test. This is defined as $(1-\beta)$ where β is the probability of the type II error. The power of a test increases as the size of the sample increases. The power tends to α as the true value θ tends to the assumed value θ_0.

Probability generating function. A function of t, usually denoted by $P(t)$, in which the probability, p_r, of r successes is the coefficient of t^r. The p.g.f. of the binomial distribution is $P(t) = (1-p+pt)^n$. The p.g.f. of the Poisson distribution is $P(t) = e^{-a+at}$. The mean and variance of the distribution are given respectively by $P'(1)$ and $P''(1)+P'(1)-\{P'(1)\}^2$.

Quartiles. See 'Lower quartile' and 'Upper quartile'.

Sampling distribution. Suppose that all the possible samples of a given size are drawn from a parent population. The means of these samples themselves form a population whose distribution is called the sampling distribution of the mean. Such a sampling distribution exists not only for the mean but for any point estimate.

Selections. See 'Combinations'.

Sheppard's correction for grouping. When the mean and variance are calculated from a grouped frequency distribution such as that of table 7A, page 67, errors occur because each observation in a group takes the mid-value of the group. The final error in the mean is negligible because the positive and negative errors in the individual observations tend to cancel each other. In the calculation of the variance, owing to squaring, all the terms become positive. An allowance can be made for the error thus caused by grouping. It is to *reduce the variance by* $\frac{1}{12}c^2$ *where c is the length of the class interval.* This is known as Sheppard's correction. The formula for the standard deviation thus becomes

$$s = \sqrt{\left\{\frac{\Sigma fx^2}{\Sigma f} - \left(\frac{\Sigma fx}{\Sigma f}\right)^2 - \frac{c^2}{12}\right\}}.$$

It will be realized that 'working units' are often chosen so that $c = 1$. The correction only applies when the group intervals are equal.

Significance. The level of significance is the *probability*, stated as α %, which is calculated when making a significance test. For some tests, by using an appropriate formula, the level of significance may be obtained from t, χ^2 or F tables. In other tests, by assuming that the null hypothesis is true, we can make a direct calculation of the probability, α %, of all values of the variate which are as extreme as, or more extreme than, the observed value. If α is greater than the *arbitrary* value 5 it is usual to accept the null hypothesis as reasonable for the population being tested by the observation or sample. If α % is less than 5 %, it is usual to reject the null hypothesis. Significance tests can be applied in a great variety of ways. They can be made more stringent by taking $2\frac{1}{2}$, 1 or 0.1 as arbitrary values for α instead of 5 but they never completely prove or disprove the null hypothesis.

Significance of a single mean. To test the hypothesis that $\mu = \mu_0$ compare

$$t = \frac{|\mu_0 - m|}{s/\sqrt{n}}$$

with the percentage points of the t-distribution given in table A5 ($\nu = n-1$).

Significance of the difference between means.

For large samples $$\frac{|m_1 - m_2|}{\sqrt{\left(\frac{s_1^2}{n_1} + \frac{s_2^2}{n_2}\right)}} > 1.96.$$

For small samples $$t = \frac{|m_1 - m_2|}{s\sqrt{\left(\frac{1}{n_1} + \frac{1}{n_2}\right)}}, \quad \text{where} \quad s^2 = \frac{(n_1-1)s_1^2 + (n_2-1)s_2^2}{(n_1+n_2-2)}.$$

Compare t with the percentage points given in table A5 taking

$$\nu = n_1 + n_2 - 2.$$

Significance of variance ratio. See 'F-test'.

Standard deviation. The standard deviation of the n values $x_1, x_2, \ldots, x_n$ is

$$s = \sqrt{\left\{\frac{\Sigma(x-\bar{x})^2}{n-1}\right\}}$$

$$= \sqrt{\left\{\frac{\Sigma x}{n-1} - \frac{n}{n-1}\left(\frac{\Sigma x}{n}\right)^2\right\}}.$$

For large samples ($n > 50$ say), $1/(n-1)$ is approximately equal to $1/n$.

If the n values have respective frequencies $f_1, f_2, \ldots, f_n$

$$s = \sqrt{\left\{\frac{\Sigma f(x-\bar{x})^2}{\Sigma f}\right\}}$$

$$= \sqrt{\left\{\frac{\Sigma f x^2}{\Sigma f} - \left(\frac{\Sigma f x}{\Sigma f}\right)^2\right\}}.$$

For a continuous probability curve $y = f(x)$ $(a \leqslant x \leqslant b)$, for which $\int_a^b y\,dx = 1$

$$\sigma = \sqrt{\left\{\int_a^b x^2 y\,dx - \left(\int_a^b xy\,dx\right)^2\right\}}.$$

Standardised deviate. The standardised deviate of a value x is $(x-\bar{x})/s$.

Type I and Type II errors. An erroneous conclusion to a significance test may be reached in two ways. We may

EITHER (i) reject the null hypothesis, H_0, when it is actually true,

OR (ii) accept the null hypothesis, H_0, when it is actually false.

These are respectively Type I and Type II errors. Their respective probabilities are always represented by α and β. Symbolically

$$\alpha = P\{\text{Type I error}\} = P\{\text{reject } H_0 | H_0 \text{ true}\},$$

$$\beta = P\{\text{Type II error}\} = P\{\text{accept } H_0 | H_0 \text{ false}\}.$$

The probability α is the level of significance of the test. It can be calculated from the value θ_0 of the parameter assumed in the null hypothesis and the data of the sample or experiment. The probability β depends not only upon θ_0 but also upon the true value θ of the parameter.

Upper quartile. The upper quartile divides the area under the probability curve (not necessarily 'normal') in the ratio 3:1. The upper quartile is the 75th percentile.

Variance ratio. See 'F-test'.

Yates's correction for continuity. In the calculation of χ^2, if $\nu = 1$ the $(O-E)$ differences must each be diminished numerically by $\frac{1}{2}$.

APPENDIX

Tables A1, 2, 4, 5, 7, 8. Reprinted from Lindley and Miller, *Cambridge Elementary Statistical Tables.*
Table A6. The e^{-x} table is taken from Godfrey and Siddons, *Four-Figure Tables.*

TABLE A1

x	y	x	y	x	y	x	y
0.0	0.3989	**1.0**	0.2420	**2.0**	0.0540	**3.0**	0.0044
.1	.3970	**.1**	.2179	**.1**	.0440	**.1**	.0033
.2	.3910	**.2**	.1942	**.2**	.0355	**.2**	.0024
.3	.3814	**.3**	.1714	**.3**	.0283	**.3**	.0017
.4	.3683	**.4**	.1497	**.4**	.0224	**.4**	.0012
0.5	0.3521	**1.5**	0.1295	**2.5**	0.0175	**3.5**	0.0009
.6	.3332	**.6**	.1109	**.6**	.0136	**.6**	.0006
.7	.3123	**.7**	.0940	**.7**	.0104	**.7**	.0004
.8	.2897	**.8**	.0790	**.8**	.0079	**.8**	.0003
.9	.2661	**.9**	.0656	**.9**	.0060	**.9**	.0002
1.0	0.2420	**2.0**	0.0540	**3.0**	0.0044	**4.0**	0.0001

Values of the ordinate $y = \frac{1}{\sqrt{(2\pi)}} e^{-\frac{1}{2}x^2}$ of the normal probability curve.

Table A2

x	A(x)		x	A(x)		x	A(x)		x	A(x)		x	A(x)		x	A(x)		x	A(x)	
0.00	0.5000	40	**0.50**	0.6915	35	**1.00**	0.8413	25	**1.50**	0.9332	13	**2.00**	0.97725	53	**2.50**	0.99379	17	**2.90**	0.99813	6
.01	.5040	40	**.51**	.6950	35	**.01**	.8438	23	**.51**	.9345	12	**.01**	.97778	53	**.51**	.99396	17	**.91**	.99819	6
.02	.5080	40	**.52**	.6985	34	**.02**	.8461	24	**.52**	.9357	13	**.02**	.97831	51	**.52**	.99413	17	**.92**	.99825	6
.03	.5120	40	**.53**	.7019	35	**.03**	.8485	23	**.53**	.9370	12	**.03**	.97882	10	**.53**	.99430	16	**.93**	.99831	5
.04	.5160	39	**.54**	.7054	34	**.04**	.8508	23	**.54**	.9382	12	**.04**	.97932	50	**.54**	.99446	15	**.94**	.99836	5
0.05	0.5199	40	**0.55**	0.7088	35	**1.05**	0.8531	23	**1.55**	0.9394	12	**2.05**	0.97982	48	**2.55**	0.99461	16	**2.95**	0.99841	5
.06	.5239	40	**.56**	.7123	34	**.06**	.8554	23	**.56**	.9406	12	**.06**	.98030	47	**.56**	.99477	15	**.96**	.99846	5
.07	.5279	40	**.57**	.7157	33	**.07**	.8577	22	**.57**	.9418	11	**.07**	.98077	47	**.57**	.99492	14	**.97**	.99851	5
.08	.5319	40	**.58**	.7190	34	**.08**	.8599	22	**.58**	.9429	12	**.08**	.98124	45	**.58**	.99506	14	**.98**	.99856	5
.09	.5359	39	**.59**	.7224	33	**.09**	.8621	22	**.59**	.9441	11	**.09**	.98169	45	**.59**	.99520	14	**.99**	.99861	4
0.10	0.5398	40	**0.60**	0.7257	34	**1.10**	0.8643	22	**1.60**	0.9452	11	**2.10**	0.98214	43	**2.60**	0.99534	13	**3.0**	0.99865	38
.11	.5438	40	**.61**	.7291	33	**.11**	.8665	21	**.61**	.9463	11	**.11**	.98257	43	**.61**	.99547	13	**3.1**	.99903	28
.12	.5478	39	**.62**	.7324	33	**.12**	.8686	22	**.62**	.9474	10	**.12**	.98300	41	**.62**	.99560	13	**3.2**	.99931	21
.13	.5517	40	**.63**	.7357	32	**.13**	.8708	21	**.63**	.9484	11	**.13**	.98341	41	**.63**	.99573	12	**3.3**	.99952	14
.14	.5557	39	**.64**	.7389	33	**.14**	.8729	20	**.64**	.9495	10	**.14**	.98392	40	**.64**	.99585	13	**3.4**	.99966	11
0.15	0.5596	40	**0.65**	0.7422	32	**1.15**	0.8749	21	**1.65**	0.9505	10	**2.15**	0.98422	39	**2.65**	0.99598	11	**3.5**	0.99977	7
.16	.5636	39	**.66**	.7454	32	**.16**	.8770	20	**.66**	.9515	10	**.16**	.98461	39	**.66**	.99609	12	**3.6**	.99984	5
.17	.5675	39	**.67**	.7486	31	**.17**	.8790	20	**.67**	.9525	10	**.17**	.98500	37	**.67**	.99621	11	**3.7**	.99989	4
.18	.5714	39	**.68**	.7517	32	**.18**	.8810	20	**.68**	.9535	10	**.18**	.98537	37	**.68**	.99632	11	**3.8**	.99993	2
.19	.5753	40	**.69**	.7549	31	**.19**	.8830	19	**.69**	.9545	9	**.19**	.98574	36	**.69**	.99643	10	**3.9**	.99995	2
0.20	0.5793	39	**0.70**	0.7580	31	**1.20**	0.8849	20	**1.70**	0.9554	10	**2.20**	0.98610	35	**2.70**	0.99653	11	**4.0**	0.99997	
.21	.5832	39	**.71**	.7611	31	**.21**	.8869	19	**.71**	.9564	9	**.21**	.98645	34	**.71**	.99664	10			
.22	.5871	39	**.72**	.7642	31	**.22**	.8888	19	**.72**	.9573	9	**.22**	.98679	34	**.72**	.99674	9			
.23	.5910	38	**.73**	.7673	31	**.23**	.8907	18	**.73**	.9582	9	**.23**	.98713	32	**.73**	.99683	10			
.24	.5948	39	**.74**	.7704	30	**.24**	.8925	19	**.74**	.9591	8	**.24**	.98745	33	**.74**	.99693	9			

x	A(x)	Δ	x	A(x)	Δ	x	A(x)	Δ	x	A(x)	Δ	x	A(x)	Δ	x	A(x)	Δ
.26	.6026	38	**.76**	.7764	30	**.26**	.8962	18	**.76**	.9608	8	**.26**	.98809	31	**.76**	.99711	9
.27	.6064	39	**.77**	.7794	29	**.27**	.8980	17	**.77**	.9616	9	**.27**	.98840	30	**.77**	.99720	8
.28	.6103	38	**.78**	.7823	29	**.28**	.8997	18	**.78**	.9625	8	**.28**	.98870	29	**.78**	.99728	8
.29	.6141	38	**.79**	.7852	29	**.29**	.9015	17	**.79**	.9633	8	**.29**	.98899	29	**.79**	.99736	8
0.30	0.6179	38	**0.80**	0.7881	29	**1.30**	0.9032	17	**1.80**	0.9641	8	**2.30**	0.98928	28	**2.80**	0.99744	8
.31	.6217	38	**.81**	.7910	29	**.31**	.9049	17	**.81**	.9649	7	**.31**	.98956	27	**.81**	.99752	8
.32	.6255	38	**.82**	.7939	28	**.32**	.9066	16	**.82**	.9656	8	**.32**	.98983	27	**.82**	.99760	7
.33	.6293	38	**.83**	.7967	28	**.33**	.9082	17	**.83**	.9664	7	**.33**	.99010	26	**.83**	.99767	7
.34	.6331	37	**.84**	.7995	28	**.34**	.9099	16	**.84**	.9671	7	**.34**	.99036	25	**.84**	.99774	7
0.35	0.6368	38	**0.85**	0.8023	28	**1.35**	0.9115	16	**1.85**	0.9678	8	**2.35**	0.99061	25	**2.85**	0.99781	7
.36	.6406	37	**.86**	.8051	27	**.36**	.9131	16	**.86**	.9686	7	**.36**	.99086	25	**.86**	.99788	7
.37	.6443	37	**.87**	.8078	28	**.37**	.9147	15	**.87**	.9693	6	**.37**	.99111	23	**.87**	.99795	6
.38	.6480	37	**.88**	.8106	27	**.38**	.9162	15	**.88**	.9699	7	**.38**	.99134	24	**.88**	.99801	6
.39	.6517	37	**.89**	.8133	26	**.39**	.9177	15	**.89**	.9706	7	**.39**	.99158	22	**.89**	.99807	6
0.40	0.6554	37	**0.90**	0.8159	27	**1.40**	0.9192	15	**1.90**	0.9713	6	**2.40**	0.99180	22	**2.90**	0.99813	
.41	.6591	37	**.91**	.8186	26	**.41**	.9207	15	**.91**	.9719	7	**.41**	.99202	22			
.42	.6628	36	**.92**	.8212	26	**.42**	.9222	14	**.92**	.9726	6	**.42**	.99224	21			
.43	.6664	36	**.93**	.8238	26	**.43**	.9236	15	**.93**	.9732	6	**.43**	.99245	21			
.44	.6700	36	**.94**	.8264	25	**.44**	.9251	14	**.94**	.9738	6	**.44**	.99266	20			
0.45	0.6736	36	**0.95**	0.8289	26	**1.45**	0.9265	14	**1.95**	0.9744	6	**2.45**	0.99286	19			
.46	.6772	36	**.96**	.8315	25	**.46**	.9279	13	**.96**	.9750	6	**.46**	.99305	19			
.47	.6808	36	**.97**	.8340	25	**.47**	.9292	14	**.97**	.9756	5	**.47**	.99324	19			
.48	.6844	35	**.98**	.8365	24	**.48**	.9306	13	**.98**	.9761	6	**.48**	.99343	18			
.49	.6879	36	**.99**	.8389	24	**.49**	.9319	13	**.99**	.9767	5	**.49**	.99361	18			
0.50	0.6915		**1.00**	0.8413		**1.50**	0.9332		**2.00**	0.9772		**2.50**	0.99379				

The function tabulated above is the total area $A(x)$ under the normal probability curve to the left of a given value of x. Mathematically it is stated as

$$A(x) = \frac{1}{\sqrt{(2\pi)}}\int_{-\infty}^{x} e^{-\frac{1}{2}t^2}\,dt.$$

TABLE A3

Rules for determining the number of degrees of freedom, ν, when applying the χ^2-test

Test of	Restrictions	No. of degrees of freedom ν
Normal distribution	Means, Standard deviations, Totals } agree	(No. of classes − 3)
Binomial distribution	(*a*) If p is given Totals agree	(No. of classes − 1)
	(*b*) If p has to be determined Means, Totals } agree	(No. of classes − 2)
Poisson distribution	(*a*) If a is given Totals agree	(No. of classes − 1)
	(*b*) If a has to be determined from the data Means, Totals } agree	(No. of classes − 2)
$1 \times k$ Contingency table	Totals agree	$(k-1)$
$h \times k$ Contingency table	h row totals agree k column totals agree	$(h-1)(k-1)$

TABLE A4

Percentage points of the χ^2-distribution

P	99.5	99	97.5	95	10	5	2.5	1	0.5	0.1
$\nu = 1$	0.0^4393	0.0^3157	0.0^3982	0.00393	2.71	3.84	5.02	6.63	7.88	10.83
2	0.0100	0.0201	0.0506	0.103	4.61	5.99	7.38	9.21	10.60	13.81
3	0.0717	0.115	0.216	0.352	6.25	7.81	9.35	11.34	12.84	16.27
4	0.207	0.297	0.484	0.711	7.78	9.49	11.14	13.28	14.86	18.47
5	0.412	0.554	0.831	1.15	9.24	11.07	12.83	15.09	16.75	20.52
6	0.676	0.872	1.24	1.64	10.64	12.59	14.45	16.81	18.55	22.46
7	0.989	1.24	1.69	2.17	12.02	14.07	16.01	18.48	20.28	24.32
8	1.34	1.65	2.18	2.73	13.36	15.51	17.53	20.09	21.95	26.12
9	1.73	2.09	2.70	3.33	14.68	16.92	19.02	21.67	23.59	27.88
10	2.16	2.56	3.25	3.94	15.99	18.31	20.48	23.21	25.19	29.59
11	2.60	3.05	3.82	4.57	17.28	19.68	21.92	24.73	26.76	31.26
12	3.07	3.57	4.40	5.23	18.55	21.03	23.34	26.22	28.30	32.91
13	3.57	4.11	5.01	5.89	19.81	22.36	24.74	27.69	29.82	34.53
14	4.07	4.66	5.63	6.57	21.06	23.68	26.12	29.14	31.32	36.12
15	4.60	5.23	6.26	7.26	22.31	25.00	27.49	30.58	32.80	37.70
16	5.14	5.81	6.91	7.96	23.54	26.30	28.85	32.00	34.27	39.25
17	5.70	6.41	7.56	8.67	24.77	27.59	30.19	33.41	35.72	40.79
18	6.26	7.01	8.23	9.39	25.99	28.87	31.53	34.81	37.16	42.31
19	6.84	7.63	8.91	10.12	27.20	30.14	32.85	36.19	38.58	43.82
20	7.43	8.26	9.59	10.85	28.41	31.41	34.17	37.57	40.00	45.31
21	8.03	8.90	10.28	11.59	29.62	32.67	35.48	38.93	41.40	46.80
22	8.64	9.54	10.98	12.34	30.81	33.92	36.78	40.29	42.80	48.27
23	9.26	10.20	11.69	13.09	32.01	35.17	38.08	41.64	44.18	49.73
24	9.89	10.86	12.40	13.85	33.20	36.42	39.36	42.98	45.56	51.18
25	10.52	11.52	13.12	14.61	34.38	37.65	40.65	44.31	46.93	52.62
26	11.16	12.20	13.84	15.38	35.56	38.89	41.92	45.64	48.29	54.05
27	11.81	12.88	14.57	16.15	36.74	40.11	43.19	46.96	49.64	55.48
28	12.46	13.56	15.31	16.93	37.92	41.34	44.46	48.28	50.99	56.89
29	13.12	14.26	16.05	17.71	39.09	42.56	45.72	49.59	52.34	58.30
30	13.79	14.95	16.79	18.49	40.26	43.77	46.98	50.89	53.67	59.70
40	20.71	22.16	24.43	26.51	51.81	55.76	59.34	63.69	66.77	73.40
50	27.99	29.71	32.36	34.76	63.17	67.50	71.42	76.15	79.49	86.66
60	35.53	37.48	40.48	43.19	74.40	79.08	83.30	88.38	91.95	99.61
70	43.28	45.44	48.76	51.74	85.53	90.53	95.02	100.4	104.2	112.3
80	51.17	53.54	57.15	60.39	96.58	101.9	106.6	112.3	116.3	124.8
90	59.20	61.75	65.65	69.13	107.6	113.1	118.1	124.1	128.3	137.2
100	67.33	70.06	74.22	77.93	118.5	124.3	129.6	135.8	140.2	149.4

The function tabulated is χ^2_P defined by the equation $\frac{P}{100} = \frac{1}{2^{\nu/2}\Gamma(\frac{1}{2}\nu)} \int_{\chi^2_P}^{\infty} x^{\frac{1}{2}\nu-1} e^{-x/2}\, dx$. If x is variable distributed as χ^2 with ν degrees of freedom, $P/100$ is the probability that $x \geqslant \chi^2_P$. For < 100, linear interpolation in ν is adequate. For $\nu > 100$, $\sqrt{(2\chi^2)}$ is approximately normally istributed with mean $\sqrt{(2\nu - 1)}$ and unit variance, and the percentage points may be obtained from able A2.

TABLE A5

Percentage points of the t-distribution

P	25	10	5	2	1	0.2	0.1	$\frac{120}{\nu}$
$\nu = 1$	2.41	6.31	12.71	31.82	63.66	318.3	636.6	
2	1.60	2.92	4.30	6.96	9.92	22.33	31.60	
3	1.42	2.35	3.18	4.54	5.84	10.21	12.92	
4	1.34	2.13	2.78	3.75	4.60	7.17	8.61	
5	1.30	2.02	2.57	3.36	4.03	5.89	6.87	
6	1.27	1.94	2.45	3.14	3.71	5.21	5.96	
7	1.25	1.89	2.36	3.00	3.50	4.79	5.41	
8	1.24	1.86	2.31	2.90	3.36	4.50	5.04	
9	1.23	1.83	2.26	2.82	3.25	4.30	4.78	
10	1.22	1.81	2.23	2.76	3.17	4.14	4.59	12
12	1.21	1.78	2.18	2.68	3.05	3.93	4.32	10
15	1.20	1.75	2.13	2.60	2.95	3.73	4.07	8
20	1.18	1.72	2.09	2.53	2.85	3.55	3.85	6
24	1.18	1.71	2.06	2.49	2.80	3.47	3.75	5
30	1.17	1.70	2.04	2.46	2.75	3.39	3.65	4
40	1.17	1.68	2.02	2.42	2.70	3.31	3.55	3
60	1.16	1.67	2.00	2.39	2.66	3.23	3.46	2
120	1.16	1.66	1.98	2.36	2.62	3.16	3.37	1
∞	1.15	1.64	1.96	2.33	2.58	3.09	3.29	0

The function tabulated is t_P defined by the equation

$$\frac{P}{100} = \frac{1}{\sqrt{(\nu\pi)}} \frac{\Gamma(\frac{1}{2}\nu+\frac{1}{2})}{\Gamma(\frac{1}{2}\nu)} \int_{|t|\geqslant t_P} \frac{dt}{(1+t^2/\nu)^{\frac{1}{2}(\nu+1)}}.$$

If t is the ratio of a random variable, normally distributed with zero mean, to an independent estimate of its standard deviation based on ν degrees of freedom, $P/100$ is the probability that $|t| \geqslant t_P$.

Interpolation ν-wise should be linear in $120/\nu$.

Other percentage points may be found approximately, except when ν and P are both small, by using the fact that the variable

$$y = \pm \sinh^{-1}\{\sqrt{(3t^2/2\nu)}\},$$

where y has the same sign as t, is approximately normally distributed with zero mean and variance $3/(2\nu-1)$.

APPENDIX

TABLE A6

e^{-x} (*for use with the Poisson distribution*)

x	.00	.01	.02	.03	.04	.05	.06	.07	.08	.09
0.0	1.0000	.9900	.9802	.9704	.9608	.9512	.9418	.9324	.9231	.9139
0.1	0.9048	.8958	.8869	.8781	.8694	.8607	.8521	.8437	.8353	.8270
.2	.8187	.8106	.8025	.7945	.7866	.7788	.7711	.7634	.7558	.7483
.3	.7408	.7334	.7261	.7189	.7118	.7047	.6977	.6907	.6839	.6771
.4	.6703	.6637	.6570	.6505	.6440	.6376	.6313	.6250	.6188	.6126
.5	.6065	.6005	.5945	.5886	.5827	.5769	.5712	.5655	.5599	.5543
.6	.5488	.5434	.5379	.5326	.5273	.5220	.5169	.5117	.5066	.5016
.7	.4966	.4916	.4868	.4819	.4771	.4724	.4677	.4630	.4584	.4538
.8	.4493	.4449	.4404	.4360	.4317	.4274	.4232	.4190	.4148	.4107
.9	.4066	.4025	.3985	.3946	.3906	.3867	.3829	.3791	.3753	.3716
1.0	0.3679	.3642	.3606	.3570	.3535	.3499	.3465	.3430	.3396	.3362
1.1	.3329	.3296	.3263	.3230	.3198	.3166	.3135	.3104	.3073	.3042
.2	.3012	.2982	.2952	.2923	.2894	.2865	.2837	.2808	.2780	.2753
.3	.2725	.2698	.2671	.2645	.2618	.2592	.2567	.2541	.2516	.2491
.4	.2466	.2441	.2417	.2393	.2369	.2346	.2322	.2299	.2276	.2254
.5	.2231	.2209	.2187	.2165	.2144	.2122	.2101	.2080	.2060	.2039
.6	.2019	.1999	.1979	.1959	.1940	.1920	.1901	.1882	.1864	.1845
.7	.1827	.1809	.1791	.1773	.1755	.1738	.1720	.1703	.1686	.1670
.8	.1653	.1637	.1620	.1604	.1588	.1572	.1557	.1541	.1526	.1511
.9	.1496	.1481	.1466	.1451	.1437	.1423	.1409	.1395	.1381	.1367
2.0	0.1353	.1340	.1327	.1313	.1300	.1287	.1275	.1262	.1249	.1237
2.1	0.1225	.1212	.1200	.1188	.1177	.1165	.1153	.1142	.1130	.1119
.2	.1108	.1097	.1086	.1075	.1065	.1054	.1044	.1033	.1023	.1013
.3	.1003	.0993	.0983	.0973	.0963	.0954	.0944	.0935	.0925	.0916
.4	.0907	.0898	.0889	.0880	.0872	.0863	.0854	.0846	.0837	.0829
.5	.0821	.0813	.0805	.0797	.0789	.0781	.0773	.0765	.0758	.0750
.6	.0743	.0735	.0728	.0721	.0714	.0707	.0699	.0693	.0686	.0679
.7	.0672	.0665	.0659	.0652	.0646	.0639	.0633	.0627	.0620	.0614
.8	.0608	.0602	.0596	.0590	.0584	.0578	.0573	.0567	.0561	.0556
.9	.0550	.0545	.0539	.0534	.0529	.0523	.0518	.0513	.0508	.0503
3.0	0.0498	.0493	.0488	.0483	.0478	.0474	.0469	.0464	.0460	.0455
3.1	.0450	.0446	.0442	.0437	.0433	.0429	.0424	.0420	.0416	.0412
.2	.0408	.0404	.0400	.0396	.0392	.0388	.0384	.0380	.0376	.0373
.3	.0369	.0365	.0362	.0358	.0354	.0351	.0347	.0344	.0340	.0337
.4	.0334	.0330	.0327	.0324	.0321	.0317	.0314	.0311	.0308	.0305
.5	.0302	.0299	.0296	.0293	.0290	.0287	.0284	.0282	.0279	.0276
.6	.0273	.0271	.0268	.0265	.0263	.0260	.0257	.0255	.0252	.0250
.7	.0247	.0245	.0242	.0240	.0238	.0235	.0233	.0231	.0228	.0226
.8	.0224	.0221	.0219	.0217	.0215	.0213	.0211	.0209	.0207	.0204
.9	.0202	.0200	.0198	.0196	.0194	.0193	.0191	.0189	.0187	.0185
4.0	0.0183									
x	.00	.01	.02	.03	.04	.05	.06	.07	.08	.09

TABLE A7

5 per cent points of the F-distribution

$\nu_1 =$	1	2	3	4	5	6	7	8	10	12	24	∞
$\nu_2 =$ **1**	161.4	199.5	215.7	224.6	230.2	234.0	236.8	238.9	241.9	243.9	249.0	254.3
2	18.5	19.0	19.2	19.2	19.3	19.3	19.4	19.4	19.4	19.4	19.5	19.5
3	10.13	9.55	9.28	9.12	9.01	8.94	8.89	8.85	8.79	8.74	8.64	8.5[illegible]
4	7.71	6.94	6.59	6.39	6.26	6.16	6.09	6.04	5.96	5.91	5.77	5.6[illegible]
5	6.61	5.79	5.41	5.19	5.05	4.95	4.88	4.82	4.74	4.68	4.53	4.3[illegible]
6	5.99	5.14	4.76	4.53	4.39	4.28	4.21	4.15	4.06	4.00	3.84	3.6[illegible]
7	5.59	4.74	4.35	4.12	3.97	3.87	3.79	3.73	3.64	3.57	3.41	3.2[illegible]
8	5.32	4.46	4.07	3.84	3.69	3.58	3.50	3.44	3.35	3.28	3.12	2.9[illegible]
9	5.12	4.26	3.86	3.63	3.48	3.37	3.29	3.23	3.14	3.07	2.90	2.7[illegible]
10	4.96	4.10	3.71	3.48	3.33	3.22	3.14	3.07	2.98	2.91	2.74	2.5[illegible]
11	4.84	3.98	3.59	3.36	3.20	3.09	3.01	2.95	2.85	2.79	2.61	2.4[illegible]
12	4.75	3.89	3.49	3.26	3.11	3.00	2.91	2.85	2.75	2.69	2.51	2.3[illegible]
13	4.67	3.81	3.41	3.18	3.03	2.92	2.83	2.77	2.67	2.60	2.42	2.2[illegible]
14	4.60	3.74	3.34	3.11	2.96	2.85	2.76	2.70	2.60	2.53	2.35	2.1[illegible]
15	4.54	3.68	3.29	3.06	2.90	2.79	2.71	2.64	2.54	2.48	2.29	2.0[illegible]
16	4.49	3.63	3.24	3.01	2.85	2.74	2.66	2.59	2.49	2.42	2.24	2.0[illegible]
17	4.45	3.59	3.20	2.96	2.81	2.70	2.61	2.55	2.45	2.38	2.19	1.9[illegible]
18	4.41	3.55	3.16	2.93	2.77	2.66	2.58	2.51	2.41	2.34	2.15	1.9[illegible]
19	4.38	3.52	3.13	2.90	2.74	2.63	2.54	2.48	2.38	2.31	2.11	1.8[illegible]
20	4.35	3.49	3.10	2.87	2.71	2.60	2.51	2.45	2.35	2.28	2.08	1.8[illegible]
21	4.32	3.47	3.07	2.84	2.68	2.57	2.49	2.42	2.32	2.25	2.05	1.8[illegible]
22	4.30	3.44	3.05	2.82	2.66	2.55	2.46	2.40	2.30	2.23	2.03	1.7[illegible]
23	4.28	3.42	3.03	2.80	2.64	2.53	2.44	2.37	2.27	2.20	2.00	1.7[illegible]
24	4.26	3.40	3.01	2.78	2.62	2.51	2.42	2.36	2.25	2.18	1.98	1.7[illegible]
25	4.24	3.39	2.99	2.76	2.60	2.49	2.40	2.34	2.24	2.16	1.96	1.7[illegible]
26	4.23	3.37	2.98	2.74	2.59	2.47	2.39	2.32	2.22	2.15	1.95	1.6[illegible]
27	4.21	3.35	2.96	2.73	2.57	2.46	2.37	2.31	2.20	2.13	1.93	1.6[illegible]
28	4.20	3.34	2.95	2.71	2.56	2.45	2.36	2.29	2.19	2.12	1.91	1.6[illegible]
29	4.18	3.33	2.93	2.70	2.55	2.43	2.35	2.28	2.18	2.10	1.90	1.6[illegible]
30	4.17	3.32	2.92	2.69	2.53	2.42	2.33	2.27	2.16	2.09	1.89	1.6[illegible]
32	4.15	3.29	2.90	2.67	2.51	2.40	2.31	2.24	2.14	2.07	1.86	1.5[illegible]
34	4.13	3.28	2.88	2.65	2.49	2.38	2.29	2.23	2.12	2.05	1.84	1.5[illegible]
36	4.11	3.26	2.87	2.63	2.48	2.36	2.28	2.21	2.11	2.03	1.82	1.5[illegible]
38	4.10	3.24	2.85	2.62	2.46	2.35	2.26	2.19	2.09	2.02	1.81	1.5[illegible]
40	4.08	3.23	2.84	2.61	2.45	2.34	2.25	2.18	2.08	2.00	1.79	1.5[illegible]
60	4.00	3.15	2.76	2.53	2.37	2.25	2.17	2.10	1.99	1.92	1.70	1.3[illegible]
120	3.92	3.07	2.68	2.45	2.29	2.18	2.09	2.02	1.91	1.83	1.61	1.2[illegible]
∞	3.84	3.00	2.60	2.37	2.21	2.10	2.01	1.94	1.83	1.75	1.52	1.0[illegible]

TABLE A8

$2\frac{1}{2}$ per cent points of the F-distribution

$\nu_1 =$	1	2	3	4	5	6	7	8	10	12	24	∞
$\nu_2 =$ 1	648	800	864	900	922	937	948	957	969	977	997	1018
2	38.5	39.0	39.2	39.2	39.3	39.3	39.4	39.4	39.4	39.4	39.5	39.5
3	17.4	16.0	15.4	15.1	14.9	14.7	14.6	14.5	14.4	14.3	14.1	13.9
4	12.22	10.65	9.98	9.60	9.36	9.20	9.07	8.98	8.84	8.75	8.51	8.26
5	10.01	8.43	7.76	7.39	7.15	6.98	6.85	6.76	6.62	6.52	6.28	6.02
6	8.81	7.26	6.60	6.23	5.99	5.82	5.70	5.60	5.46	5.37	5.12	4.85
7	8.07	6.54	5.89	5.52	5.29	5.12	4.99	4.90	4.76	4.67	4.42	4.14
8	7.57	6.06	5.42	5.05	4.82	4.65	4.53	4.43	4.30	4.20	3.95	3.67
9	7.21	5.71	5.08	4.72	4.48	4.32	4.20	4.10	3.96	3.87	3.61	3.33
10	6.94	5.46	4.83	4.47	4.24	4.07	3.95	3.85	3.72	3.62	3.37	3.08
11	6.72	5.26	4.63	4.28	4.04	3.88	3.76	3.66	3.53	3.43	3.17	2.88
12	6.55	5.10	4.47	4.12	3.89	3.73	3.61	3.51	3.37	3.28	3.02	2.72
13	6.41	4.97	4.35	4.00	3.77	3.60	3.48	3.39	3.25	3.15	2.89	2.60
14	6.30	4.86	4.24	3.89	3.66	3.50	3.38	3.29	3.15	3.05	2.79	2.49
15	6.20	4.76	4.15	3.80	3.58	3.41	3.29	3.20	3.06	2.96	2.70	2.40
16	6.12	4.69	4.08	3.73	3.50	3.34	3.22	3.12	2.99	2.89	2.63	2.32
17	6.04	4.62	4.01	3.66	3.44	3.28	3.16	3.06	2.92	2.82	2.56	2.25
18	5.98	4.56	3.95	3.61	3.38	3.22	3.10	3.01	2.87	2.77	2.50	2.19
19	5.92	4.51	3.90	3.56	3.33	3.17	3.05	2.96	2.82	2.72	2.45	2.13
20	5.87	4.46	3.86	3.51	3.29	3.13	3.01	2.91	2.77	2.68	2.41	2.09
21	5.83	4.42	3.82	3.48	3.25	3.09	2.97	2.87	2.73	2.64	2.37	2.04
22	5.79	4.38	3.78	3.44	3.22	3.05	2.93	2.84	2.70	2.60	2.33	2.00
23	5.75	4.35	3.75	3.41	3.18	3.02	2.90	2.81	2.67	2.57	2.30	1.97
24	5.72	4.32	3.72	3.38	3.15	2.99	2.87	2.78	2.64	2.54	2.27	1.94
25	5.69	4.29	3.69	3.35	3.13	2.97	2.85	2.75	2.61	2.51	2.24	1.91
26	5.66	4.27	3.67	3.33	3.10	2.94	2.82	2.73	2.59	2.49	2.22	1.88
27	5.63	4.24	3.65	3.31	3.08	2.92	2.80	2.71	2.57	2.47	2.19	1.85
28	5.61	4.22	3.63	3.29	3.06	2.90	2.78	2.69	2.55	2.45	2.17	1.83
29	5.59	4.20	3.61	3.27	3.04	2.88	2.76	2.67	2.53	2.43	2.15	1.81
30	5.57	4.18	3.59	3.25	3.03	2.87	2.75	2.65	2.51	2.41	2.14	1.79
32	5.53	4.15	3.56	3.22	3.00	2.84	2.72	2.62	2.48	2.38	2.10	1.75
34	5.50	4.12	3.53	3.19	2.97	2.81	2.69	2.59	2.45	2.35	2.08	1.72
36	5.47	4.09	3.51	3.17	2.94	2.79	2.66	2.57	2.43	2.33	2.05	1.69
38	5.45	4.07	3.48	3.15	2.92	2.76	2.64	2.55	2.41	2.31	2.03	1.66
40	5.42	4.05	3.46	3.13	2.90	2.74	2.62	2.53	2.39	2.29	2.01	1.64
60	5.29	3.93	3.34	3.01	2.79	2.63	2.51	2.41	2.27	2.17	1.88	1.48
120	5.15	3.80	3.23	2.89	2.67	2.52	2.39	2.30	2.16	2.05	1.76	1.31
∞	5.02	3.69	3.12	2.79	2.57	2.41	2.29	2.19	2.05	1.94	1.64	1.00

INDEX

The numbers refer to pages

INDEX

Printed in the United States
By Bookmasters